NEW MEDIA IN TIMES OF CRISIS

New Media in Times of Crisis provides an interdisciplinary look at research focused around how people organize during crises.

Contributors examine the latest practices for communicating during crises, including evacuation practices, workplace safety challenges, crisis social media usage, and strategies for making emergency alerts on U.S. mobile phones constructive and helpful. The book is grounded in the practices of first responders, crisis communicators, people experiencing tragic events, and communities who organize on- and offline to make sense of their experiences. The authors draw upon a wide range of theories and frameworks with the goal of establishing new directions for research and practice.

The text is suitable for advanced students and researchers in crisis, disaster, and emergency communication.

Keri K. Stephens is an Associate Professor in the Moody College of Communication, and a Distinguished Teaching Professor at The University of Texas at Austin. Her research and teaching interests bring an organizational perspective to understanding how people interact with communication technologies. She has more than 60 peer-reviewed publications, including a recent book *Negotiating Control: Organizations and Mobile Communication*.

New Agendas in Communication
A Series from Routledge and the College of Communication at the University of Texas at Austin
Karin Wilkins, Series Editor

This series brings together groups of emerging scholars to tackle important interdisciplinary themes that demand new scholarly attention and reach broadly across the communication field's existing courses. Each volume stakes out a key area, presents original findings, and considers the long-range implications of its "new agenda."

Recent series titles include:

New Technologies and Civic Engagement
Edited by Homero Gil de Zúñiga

Networked China
Global Dynamics of Digital Media and Civic Engagement
Edited by Wenhong Chen and Stephen D. Reese

Strategic Communication
Edited by Anthony Dudo and LeeAnn Kahlor

Work Pressures
Edited by Dawna I. Ballard and Matthew S. McGlone

Digital Discussions
How Big Data Informs Political Communication
Edited by Natalie Jomini Stroud and Shannon McGregor

New Media in Times of Crisis
Edited by Keri K. Stephens

For a full list of titles please visit: https://www.routledge.com/New-Agendas-in-Communication-Series/book-series/NEWAGENDAS.

NEW MEDIA IN TIMES OF CRISIS

Edited by Keri K. Stephens

Routledge
Taylor & Francis Group

NEW YORK AND LONDON

First edition published 2019
by Routledge
52 Vanderbilt Avenue, New York, NY 10017

and by Routledge
2 Park Square, Milton Park, Abingdon, Oxon, OX14 4RN

Routledge is an imprint of the Taylor & Francis Group, an informa business

Library of Congress Cataloging-in-Publication Data
A catalog record has been requested for this book

ISBN: 978-1-138-57028-3 (hbk)
ISBN: 978-1-138-57029-0 (pbk)
ISBN: 978-0-203-70363-2 (ebk)

Typeset in Bembo
by codeMantra

CONTENTS

ACKNOWLEDGMENTS

I still remember waking up early on a Saturday morning to wrap up our meeting around this book. I came into the workspace an hour early to review the notes from the day before and to write our organizing framework on a wall-length whiteboard. We represented a host of different disciplines ranging from organizational communication and public relations to computer science and civil engineering. I was not sure if we could find the crystalizing concepts needed to glue together our individual contributions. What happened that morning was inspiring. We grabbed markers and began mapping our ideas—both empirical and conceptual—all over the board. We realized that although we refer to concepts using different terms and we cite different scholarship, our shared interests were around organizing and pushing the boundaries of understanding crises, emergencies, and disasters. This book is what we have created together.

I am grateful to the Moody College of Communication for providing the funds that allowed our diverse group to assemble and create this contribution. This assembly included practitioners and emergency responders—Cindy Posey, Chief David Carter, and Charlie Moore—as well as graduate students who facilitated the sessions—Yaguang Zhu, Brett Robertson, Millie Harrison, and Ryan Crace—and Associate Dean Karin Wilkins who supported our vision.

I am grateful for my colleagues Dawna Ballard, Josh Barbour, Barry Brummett, Johanna Hartelius, Sharon Jarvis, Matt McGlone, Talia Stroud, and Jeff Treem, who shared their book-editing wisdom with me and supported this endeavor. The initial ideas for this book were shaped in part by scholars who have molded my thinking and encouraged me in my pursuit of this line of research, and those include: Chris Bailey, Larry Browning, Timothy Coombs, Laurie Lewis, Marya Doerfel, Sherry Holladay, Leysia Palen, Dan O'Hair,

Patty Malone, Amy Schmisseur, Jan-Oddvar Sørnes, Jeannette Sutton, and Scott D'Urso.

Finally, I would like to thank my family, Tab, Sarah Kay, and Kyle, for allowing me to travel, work on weekends, and pursue the type of work that makes me feel like I am making a difference in the world.

INTRODUCTION

Keri K. Stephens

Fires can level cities and oil spills threaten the environment. School shootings are too frequent and shake the confidence of the public. Sometimes there is advance warning when tragedy strikes, but often these events are unpredictable, and they always generate high levels of uncertainty. In these times of crisis, communication matters. The focus of this book is to understand human behavior—specifically organizing—now that people have so many forms of new media to use when communicating.

This book, *New Media in Times of Crisis*, uses the terms *crisis* and *new media* broadly, with the hope they can transcend disciplinary boundaries. Here, *crisis* means a time of danger, and by using a generic definition, we can employ related words such as *disasters*, *emergencies*, and *hazards* to all reflect dangerous situations. While different disciplines try to differentiate between these related concepts (e.g., disaster sociology uses different terms than public relations), crises are often considered a form of emergency (Reynolds & Seeger, 2005; Seeger, 2006), and disasters almost always constitute an emergency. But what all these events have in common is the high level of *uncertainty* that participants experience, and their need for information that is accurate and timely (Coombs, 2012; Quarantelli, 1998; Seeger, 2006; Sorensen, 2000). These situations can be life-threatening. Their unpredictable nature necessitates that emergency responders and citizens engage in countless contingency options in crisis preparation because crises rarely play out like we expect. Simply put, crises are complex (Weick & Sutcliffe, 2015).

A big reason for the complexity of crises is the vast range of communication options and technology available today. We use the term *new media* to signal this complexity and to capture the ever-changing nature of information and communication technologies (ICTs). From mobile devices to social media, people can communicate in myriad ways during crises; they can call and text another individual, or they can post a message on social media for anyone to see. They can download apps on their mobile devices and participate in social conversations on the go. Further complicating the matter, new tools and technologies are arriving on the scene all the time.

In this book, our scholars represent a host of disciplines—organizational communication, public relations, computer science, civil engineering—therefore, they will use terms like *crisis* and *new media* in slightly different ways. We view this as a strength in our research because we can begin to build a common vocabulary that allows us to collaborate on truly interdisciplinary projects. Engaging these topics through an interdisciplinary lens is not only our goal, it was our mode of operation. These authors met in person to challenge one another to think more broadly about the research they each do. During these meetings, we invited crisis communication practitioners and first responders to participate in our dialog and help us shape this contribution. We kept their messages in mind as we crafted our chapters. Their perspectives were so important to our mission that we have included some of those practitioners in our work.

Interdisciplinary is another term often used in disparate ways, and debates continue over whether we should strive for our research to be multidisciplinary, transdisciplinary, or interdisciplinary (e.g., Choi & Pak, 2006). We have followed the lead of the National Science Foundation (n.d.) in striving to approach these topics by including scholars trained in diverse disciplines. Crises transcend the scope of individual fields, and together we all share a desire to understand how to approach topics like organizing and communicating in novel ways. It is noteworthy that despite our disciplinary differences, some foundational research is cited by multiple authors in this book. For example, authors from the civil engineering team, the university crisis team, and the wireless emergency alert chapters in this book all cite some of the public warning literature that has sociological roots (e.g., Mileti & Sorensen, 1990). Another connection is in the work of Norris and colleagues and the notion of resilience (Norris, Stevens, Pfefferbaum, Wyche, & Pfefferbaum, 2008). The resilience literature is inherently interdisciplinary since that work originated in psychology, has roots in medicine, and is now funded by the Department of Homeland Security. Arguably, some of the needs articulated in Chapters 1–3 for more focus on helping emergency and crisis responders cope with the ongoing stress of their jobs could benefit from this body of literature on resilience.

Before delving into the structure of this book, let me share some foundational concepts and past research that should ground and situate this contribution.

Messages

Past researchers have written books and a host of articles concerning how to construct messages that accomplish communication goals during times of crisis (e.g., Benoit, 1997; Coombs, 1995, 2007, 2010, 2012; Coombs & Holladay, 2011;

Mileti & Sorensen, 1990; Seeger & Griffin Padgett, 2010; Stephens & Barrett, 2016; Stephens & Malone, 2009; Stephens, Malone, & Bailey, 2005). Coombs (2007; Coombs & Holladay, 2002) developed the well-cited Situational Crisis Communication Theory (SCCT), which is discussed in detail in Chapter 5 of this book. Scholars also have integrated concepts in SCCT with rhetorical strategies, and many practitioners have relied on this well-researched advice concerning how to communicate with their publics. There are handbooks written on warnings (Wogalter, 2006) that comprehensively discuss how and when to send messages in a host of situations. Notably, one of the most foundational understandings of how people will interpret emergency messages is found in Mileti and Sorensen's (1990) model of "hear-confirm-understand-decide-respond." This is discussed in detail in Chapter 6 where Bean and Madden relate this to an understanding of wireless emergency alerts (WEAs).

Technology

Along with message strategies, researchers have studied the evolution of various ICTs used to disseminate crisis and emergency messages. The earlier research focused on using a mix of ICTs (e.g., González-Herrero & Smith, 2008; Palen, Hiltz, & Liu, 2007; Procopio & Procopio, 2007; Schultz, Utz, & Göritz, 2011; Stephens, Barrett, & Mahometa, 2013; Sweetser & Metzgar, 2007). As can be imagined, this research area exploded with the emergence and widespread use of social media. Initially these studies focused on what were called microblogs (e.g., Heverin & Zach, 2012; Vieweg, Hughes, Starbird, & Palen, 2010)—tools that evolved into Twitter. Since then researchers have attempted to understand how messages sent through Twitter might function in disasters and crises (e.g., Li & Rao, 2010; Marwick & boyd, 2011; Peary, Shaw, & Takeuchi, 2012; Utz, Schultz, & Glocka, 2012). There were articles examining how social media were being incorporated into risk and crisis communication (Veil, Buehner, & Palenchar, 2011), and how they were changing the practice of public relations (Wright & Hinson, 2008).

Two helpful perspectives emerged from this research that have informed the most current thinking on new media use in crises. First, the field of crisis informatics established an interdisciplinary perspective integrating computer science and human behavior. This body of research (e.g., Hughes & Palen, 2009; Hughes, Palen, Sutton, Liu, & Vieweg, 2008; Palen et al., 2010; Palen, Vieweg, & Liu, 2009; Starbird & Palen, 2010; Sutton, Palen, & Shklovski, 2008) revealed the power that individuals have at their fingertips because they can instantaneously post messages publicly using social media. Their research articulated the importance of timeliness in responses as well how online convergence of disparate groups—the topic of Chapters 9 and 10 in this book—happens. Second, and most recently, Austin, Liu, and Jin (2012) developed a model of social-mediated crisis communication. This research prompted the publication of *Social Media and Crisis Communication* (Austin & Jin, 2018).

Temporal Considerations in Times of Crisis

An undercurrent to much of the past research has focused on segmenting the study of crises and emergencies into phases, most often pre-, during, and post-crisis. Planning is typically the focus in the pre-crisis phase, and this research engages topics of leadership (Fink, 1986; Mitroff, 2004) as well as the importance of having strong relationships before a negative event occurs (Coombs, 2012; Ulmer, 2001). In this book, Chapter 7 discusses transportation planning and how that can impact evacuations. Research during a crisis often focuses on messaging, as discussed in the previous section of this chapter. Now scholars can harness big data to capture a corpus of public Twitter posts and run automated analyses like the authors of Chapter 5 in this book do. Post-crisis research acknowledges that ending points can be fuzzy, and that immediately after the threat has subsided, there can still be extensive rebuilding of physical structures (in the case of a disaster) as well as emotional and reputational concerns. Seeger and Griffin Padgett (2010) have developed a theoretical guide to understanding the renewal process. In addition, there has been research on how communities and individuals can develop resilience to help them recover, and if necessary, prepare for future tragic events (Norris et al., 2008). Community resilience is the topic of Chapter 8 of this book.

While it is often convenient to segment crises into phases, there is also a clear acknowledgment that crises, emergencies, and disasters, should be viewed as cyclical. For example, Reynolds and Seeger (2005) encouraged crisis planners to include what they call the full cycle of communication practices, that encompasses pre-crisis, during crisis, and post-crisis actions. In disasters and emergencies there are often after-action reports that explore what happened and how the involved parties can adapt and learn from their experiences (Ishak & Williams, 2017). Chapter 1 of this book examines how a university crisis communication team has used these after-action reports to learn and adapt to the constantly changing new media environment.

Beyond Phases Toward an Organizing Focus

Past research has laid the groundwork for this book to build on these ideas by capitalizing on our interdisciplinary perspectives of organizing and human behavior in this new media environment. The biggest change, from an organizational perspective, that necessitates this transformation in thinking is how control over messages and media has shifted from formal organizations, to be more distributed and include the public (Stephens & Ford, 2014). Now, organizations—for example, companies or emergency response groups—can plan their messages and be strategic, but the public can weigh in and modify the messaging. People carry their personal mobile devices with them everywhere to stay connected and easily join conversations (Bayer, Campbell, & Ling, 2016;

Campbell & Park, 2008; Ling & Yttri, 2002; Stephens, 2018). Control is often an illusion; organizations must expand their practices to understand how their publics draw upon the affordances of new media to organize among themselves, co-creating crisis messages and planning their own interventions. For example, the devastating hurricanes in the U.S. during 2017 and 2018 clearly illustrated how citizen groups organized to rescue individuals because official rescue resources were overloaded (Smith, Stephens, Robertson, Li, & Murthy, 2018). Finally, first responders and their managers also need to understand how organizing practices can help them garner resources as well as influence how they interact with the public.

Barbour, Buzzanell, Kinsella, and Stephens (2018) assembled a group of scholars all interested in communicating for reliability, resilience, and safety and issued a call for expanding this type of research in a special issue of *Corporate Communications: An International Journal*. This organizing-focused approach is also seen the work of Doerfel and her colleagues (Chewning, Lai, & Doerfel, 2013; Doerfel, Lai, & Chewning, 2009) as they explored interorganizational relationships as well as the role ICTs played in helping rebuild communication structures post-disaster. Chapter 3 in this book focuses on interorganizational relationships that Williams calls multiteams. Our work in this book further extends this trend and invites even broader perspectives to approach these topics in the most interdisciplinary perspective to date.

Overview of This Book

Starting in the first chapter of this book, we introduce our focus on organizing (i.e. processes), rather than organizations (i.e., structure) by relying on the foundational work of Weick (1979). He asserts that organizations are in a constant state of change, and that is clearly seen in many chapters of this book. Another idea conceptualized and explored by Weick (1993, 1995) is the notion of sensemaking, the process where people retrospectively, and often collectively, formulate an explanation of what they experienced. This concept is explored with considerable depth in Chapter 2 when Jahn elaborates on different ways that firefighters engage in sensemaking around close calls. Sensemaking and related topics appear in many of the other chapters as well. Figure 0.1 illustrates how the various past research, along with the chapters in this book can be seen as subsets of organizing.

The figure shows the invasive nature of organizing processes even though we can clearly see that there are times that a focus on new media, messages, crisis responders, or public involvement is possible. Section I of this book begins with a look at first responders and crisis communicators. In the preview to that section, I will share some of the history of social and mobile media to set the stage for why practitioners are being asked to learn and change their practices at an extreme pace. This is important context for Chapter 1 where Barrett and

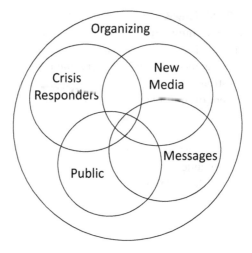

FIGURE 0.1 Model of Organizing in Times of Crisis (OTC)

Posey examine several different campus emergencies, the role new media have played, the importance of understanding how the public participates in crisis storytelling. By the end of Chapter 3, readers will better understand organizing as a process.

Section II of this book focuses more on how various individuals and publics seek and share crisis information. From workplace safety and WEAs to angry Twitter posts and the challenges of evacuations, these chapters help readers see the individual's role in crisis communication. The individuals' role in organizing practices is important because, ultimately, people decide for themselves the actions they will take. In Chapter 4, Ford examines safety information seeking at work, and concepts related to risk communication and the safety cultures organizations try to create. Cacciatore, Kim, and Danzy demonstrate in Chapter 5 analysis techniques that characterize the types of messages individuals posted during an United Airlines crisis. This showcases the role individual members of the public plays as crises unfold. This is followed by Bean and Madden's Chapter 6 exploring WEAs, and whether including maps in these brief alerts could help people make sense of these messages. Finally, Rambha, Jafari, and Boyles provide a comprehensive review that addresses individuals' evacuation and sheltering behaviors in Chapter 7.

The final section of the book explores different ways that groups organize to cope, collaborate, and create new organizations with lives of their own. Houston's Chapter 8 provides a state-of-the-art review of the community resilience literature. In Chapter 9, Hughes, one of the original researchers in the crisis informatics field, shows how advances in new media have allowed groups to converge online in more profound ways. Lai, in Chapter 10, extends our understanding of online convergence by developing a model describing what happens when a crisis is over and the online groups become dormant. Finally, Chapter 11 further elaborates on the Model of Organizing in Times of Crisis (OTC) and integrates the chapters to create a clear interdisciplinary agenda for future research.

References

Austin, L., & Jin, Y. (Eds.). (2017). *Social media and crisis communication*. New York, NY: Routledge.

Austin, L., Liu, B. F., & Jin, Y. (2012). How audiences seek out crisis information: Exploring the social-mediated crisis communication model. *Journal of Applied Communication Research, 40*, 188–207. https://doi.org/10.1080/00909882.2012.654498

Barbour, J. B., Buzzanell, P. M., Kinsella, W. J., & Stephens, K. K. (2018). Communicating/organizing for reliability, resilience, and safety: Special issue introduction. *Corporate Communications: An International Journal, 23*, 154–161. https://doi.org/10.1108/CCIJ-01-2018-0019

Bayer, J. B., Campbell, S. W., & Ling, R. (2016). Connection cues: Activating the norms and habits of social connectedness. *Communication Theory, 26*, 128–149. https://doi.org/10.1111/comt.12090

Benoit, W. L. (1997). Image repair discourse and crisis communication. *Public Relations Review, 23*, 177–186. https://doi.org/10.1016/S0363-8111(97)90023-0

Campbell, S. W., & Park, Y. J. (2008). Social implications of mobile telephony: The rise of personal communication society. *Sociology Compass, 2*, 321–387. https://doi:10.1111/j.1751-9020.2007.00080.x

Chewning, L. V., Lai, C. H., & Doerfel, M. L. (2013). Organizational resilience and using information and communication technologies to rebuild communication structures. *Management Communication Quarterly, 27*, 237–263. https://doi.org/10.1177/0893318912465815

Choi, B. C., & Pak, A. W. (2006). Multidisciplinarity, interdisciplinarity and transdisciplinarity in health research, services, education and policy: 1. Definitions, objectives, and evidence of effectiveness. *Clinical and Investigative Medicine, 6*, 351–364. https://pdfs.semanticscholar.org/7cc4/ac195c59a8fbbff463ad10f12167d9254bce.pdf

Coombs, W. T. (1995). Choosing the right words: The development of guidelines for the selection of the "appropriate" crisis-response strategies. *Management Communication Quarterly, 8*, 447–476. https://doi.org/10.1177/0893318995008004003

Coombs, W. T. (2007). Protecting organization reputations during a crisis: The development and application of Situational Crisis Communication Theory. *Corporate Reputation Review, 10*(3), 163–176. https://doi.org/10.1057/palgrave.crr.1550049

Coombs, W. T. (2010). Crisis communication: A developing field. In R. L. Heath (Ed.), *The Sage handbook of public relations* (pp. 477–488). Thousand Oaks, CA: Sage.

Coombs, W. T. (2012). *Ongoing crisis communication: Planning, managing, and responding* (2nd ed.). Thousand Oaks, CA: Sage.

Coombs, W. T., & Holladay, S. J. (2002). Helping crisis managers protect reputational assets: Initial tests of the Situational Crisis Communication Theory. *Management Communication Quarterly, 16*, 165–186. https://doi.org/10.1177/089331802237233

Coombs, W. T., & Holladay, S. J. (Eds.). (2011). *The handbook of crisis communication*. Malden, MA: John Wiley & Sons.

Doerfel, M. L., Lai, C. H., & Chewning, L. V. (2009). The evolutionary role of interorganizational communication: Modeling social capital in disaster contexts. *Human Communication Research, 36*, 125–162. https://doi.org/10.1111/j.1468-2958.2010.01371.x

Fink, S. (1986). Crisis management: Planning for the inevitable. New York: AMACOM.

González-Herrero, A., & Smith, S. (2008). Crisis communications management on the web: How Internet based technologies are changing the way public relations professionals handle business crises. *Journal of Contingencies and Crisis Management, 16*, 143–153. https://doi.org/10.1111/j.1468-5973.2008.00543.x

Heverin, T., & Zach, L. (2012). Use of microblogging for collective sense-making during violent crises: A study of three campus shootings. *Journal of the American Society of Information Science and Technology, 63*, 34–47. https://doi.org/10.1002/asi.21685

Hughes, A. L., & Palen, L. (2009). Twitter adoption and use in mass convergence and emergency events. *International Journal of Emergency Management, 6*, 248–260. https://doi.org/10.1504/IJEM.2009.031564

Hughes, A. L., Palen, L., Sutton, J., Liu, S. B., & Vieweg, S. (2008, May). Site-seeing in disaster: An examination of on-line social convergence. In *Proceedings of the 5th International ISCRAM Conference*. Washington, DC.

Ishak, A. W., & Williams, E. A. (2017). Slides in the tray: How fire crews enable members to borrow experiences. *Small Group Research, 48*(3), 336–364. https://doi.org/10.1177/1046496417697148

Li, J., & Rao, H. R. (2010). Twitter as a rapid response news service: An exploration in the context of the 2008 China earthquake. *The Electronic Journal on Information Systems in Developing Countries, 42*, 1–22. https://doi.org/10.1002/j.1681-4835.2010.tb00300.x

Ling, R., & Yttri, B. (2002). Hyper-coordination via mobile phones in Norway. In J. E. Katz & M. Aakhus (Eds.), *Perpetual contact: Mobile communication, private talk, public performance* (pp. 139–169). Cambridge: Cambridge University Press.

Marwick, A. E., & boyd, d. (2011). I tweet honestly, I tweet passionately: Twitter users, context collapse, and the imagined audience. *New Media & Society, 13*, 114–133.

Mileti, D. S., & Sorensen, J. H. (1990). *Communication of emergency public warnings: A social science perspective and state-of-the-art assessment*. Oakridge, TN: Oak Ridge National Laboratory.

Mitroff, I. I. (2004). Crisis leadership: Planning for the unthinkable. Hoboken, NJ: Wiley.

National Science Foundation. (n.d.) Introduction to interdisciplinary research. *National Science Foundation*. Retrieved from: www.nsf.gov/od/oia/additional_resources/interdisciplinary_research/

Norris, F. H., Stevens, S. P., Pfefferbaum, B., Wyche, K. F., & Pfefferbaum, R. L. (2008). Community resilience as a metaphor, theory, set of capacities, and strategy for disaster readiness. *American Journal of Community Psychology, 41*, 127–150.

Palen, L., Anderson, K. M., Mark, G., Martin, J., Sicker, D., Palmer, M., & Grunwald, D. (2010). A vision for technology-mediated support for public participation & assistance in mass emergencies & disasters. In *Proceedings of the 2010 ACM-BCS Visions of Computer Science Conference* (p. 8). British Computer Society, Edinburgh, United Kingdom.

Palen, L., Hiltz, S. R., & Liu, S. B. (2007). Online forums supporting grassroots participation in emergency preparedness and response. *Communications of the ACM, 50*, 54–58. https://doi.org/10.1145/1226736.1226766

Palen, L., Vieweg, S., & Liu, S. B. (2009). Crisis in a networked world: Features of computer-mediated communication in the April 16, 2007 Virginia Tech event. *Social Science Computer Review, 27*, 467–480. https://doi.org/10.1177/0894439309332302

Peary, B. D., Shaw, R., & Takeuchi, Y. (2012). Utilization of social media in the East Japan earthquake and tsunami and its effectiveness. *Journal of Natural Disaster Science, 34*, 3–18. https://doi.org/10.2328/jnds.34.3

Procopio, C. H., & Procopio, S. T. (2007). Do you know what it means to miss New Orleans? Internet communication, geographic community, and social capital in crisis. *Journal of Applied Communication Research, 35*, 67–87. https://doi.org/10.1080/00909880601065722

Quarantelli, E. L. (Ed.), (1998). *What is a disaster: Perspectives on the question*. London: Routledge.

Reynolds, B., & Seeger, M. W. (2005). Crisis and emergency risk communication as an integrated model. *Journal of Health Communication, 10*, 43–55. https://doi.org/10.1080/10810730590904571

Schultz, F., Utz, S., &, Göritz, A. (2011). Is the medium the message? Perceptions of and reactions to crisis communication via Twitter, blogs, and traditional media. *Public Relations Review, 37*, 20–27. https://doi.org/10.1016/j.pubrev.2010.12.001

Seeger, M. W. (2006). Best practices in crisis communication: An expert panel process. *Journal of Applied Communication, 34*, 232–244. https://doi.org/10.1080/00909880600769944

Seeger, M. W., & Griffin Padgett, D. R. (2010). From image restoration to renewal: Approaches to understanding post-crisis communication. *Review of Communication, 10*, 127–141. https://doi.org/10.1080/15358590903545263

Smith, W. R., Stephens, K. K., Robertson, B. W., Li. J., & Murthy, D. (2018). Social media in citizen-led disaster response: Rescuer roles, coordination challenges, and untapped potential. In K. Boersma & B. Tomaszewski (Eds.), *Proceedings of the 15th International ISCRAM Conference*. Rochester, NY. Retrieved from: http://idl.iscram.org/files/williamrsmith/2018/1586_WilliamR.Smith_etal2018.pdf

Sorensen, J. H. (2000). Hazard warning systems: Review of 20 years of progress. *Natural Hazards Review, 1*, 119–125. https://doi.org/10.1061/(ASCE)1527-6988(2000)1:2(119)

Starbird, K., & Palen, L. (2010). Pass it on? Retweeting in mass emergency. *Proceedings of the 7th International ISCRAM conference*. Seattle, WA. www.researchgate.net/profile/Leysia_Palen2/publication/228512367_Pass_It_On_Retweeting_in_Mass_Emergency/links/00b7d52bc84dca2d2f000000.pdf

Stephens, K. K. (2018). *Negotiating control: Organizations and mobile communication*. New York, NY: Oxford University Press.

Stephens, K. K., & Barrett, A. K. (2016). Communicating briefly: Technically. *Journal of Business Communication, 53*, 398–418. https://doi.org/10.1177/2329488414525463

Stephens, K. K., Barrett, A. K., & Mahometa, M. L. (2013). Organizational communication in emergencies: Using multiple channels and sources to combat noise and capture attention. *Human Communication Research, 39*, 230–251. https://doi.org/10.1111/hcre.12002

Stephens, K. K., & Ford, J. L. (2014). *Crisis communications and sharing message control*. In M. Khosrow-Pour (Ed.), *Encyclopedia of information science and tech* (pp. 462–470). New York: IGI Global.

Stephens, K. K., & Malone, P. C. (2009). If the organizations won't give us information... The use of multiple new media for crisis technical translation and dialogue. *Journal of Public Relations Research, 21*, 229–239. https://dx.doi.org/10.1080/10627260802557605

Stephens, K. K., Malone, P. C., & Bailey, C. (2005). Communicating with stakeholders during a crisis: Evaluating message strategies. *Journal of Business Communication, 42*, 390–419. https://doi.org/10.1177/0021943605279057

Sutton, J., Palen, L., & Shklovski, I. (2008). Backchannels on the front lines: Emergent use of social media in the 2007 Southern California fire. *Proceedings of Information Systems for Crisis Response and Management Conference (ISCRAM)*. Washington, DC. Retrieved from: www.iscram.org/legacy/dmdocuments/ISCRAM2008/papers/ISCRAM2008_Sutton_etal.pdf

Sweetser, K. D., & Metzgar, E. (2007). Communicating during crisis: Use of blogs as a relationship management tool. *Public Relations Review, 33*, 340–342. https://doi.org/10.1016/j.pubrev.2007.05.016

Ulmer, R. R. (2001). Effective crisis management through established stakeholder relationships. *Management Communication Quarterly, 14*, 590–615. https://doi.org/10.1177/0893318901144003

Utz, S., Schultz, F., & Glocka, S. (2012). Crisis communication online: How medium, crisis type and emotions affected public reactions in the Fukushima Daiichi nuclear disaster. *Public Relations Review, 39*, 40–46. https://doi.org/10.1016/j.pubrev.2012.09.010

Veil, S. R., Buehner, T., & Palenchar, M. J. (2011). A work-in-process literature review: Incorporating social media in risk and crisis communication. *Journal of Contingencies and Crisis Management, 19*, 110–122. https://doi.org/10.1111/j.1468-5973.2011.00639.x

Vieweg, S., Hughes, A. L., Starbird, K., & Palen, L. (2010). Microblogging during two natural hazards events: What Twitter may contribute to situational awareness. In *Proceedings of the SIGCHI Conference on Human Factors in Computing Systems* (pp. 1079–1088). https://doi.org/10.1145/1753326.1753486

Weick, K. E. (1979). *Social psychology of organizing*. Reading, MA: Addison-Wesley.

Weick, K. E. (1993). The collapse of sensemaking in organizations: The Mann Gulch disaster. *Administrative Science Quarterly, 38*(4), 628–652. www.jstor.org/stable/2393339

Weick, K. E. (1995). *Sensemaking in organizations* (Vol. 3). Thousand Oaks, CA: Sage.

Weick, K. E., & Sutcliffe, K. M. (2015). *Managing the unexpected: Sustained performance in a complex world* (3rd ed.). Hoboken, NJ: John Wiley & Sons, Inc.

Wogalter, M. S. (Ed.). (2006). The handbook of warnings. Mahwah, NJ: Lawrence Erlbaum.

Wright, D. K., & Hinson, M. D. (2008). How blogs and social media are changing public relations and the way it is practiced. *Public Relations Journal, 2*, 1–21. Retrieved from: http://apps.prsa.org/SearchResults/view/6D-020203/0/How_Blogs_and_Social_Media_are_Changing_Public_Rel#.W89uwPZOmUk

SECTION I
Focusing on Crisis Responders

Section I of this book focuses on a group of people who play a vital role in saving lives: crisis communication practitioners and emergency responders. Whether they are in the field saving people during disasters or working behind the scenes communicating through social media, we begin this interdisciplinary exploration by weaving in their experiences with our academic understandings.

As academics representing diverse disciplines, our research findings often coalesce around the notion of sensemaking—how people give meaning to their actions based on collective understandings (Weick, 1979). While this is a theme throughout this book, it plays a crystalizing role as we examine what can be considered team-sensemaking in-the-moment.

FIGURE 0.2 Crisis responders
Photo courtesy of Houston Police Department

In the three chapters that follow we look behind the scenes at how a university crisis communication team, a group of firefighters, and teams of emergency responders do their jobs during quickly unfolding and highly uncertain times. All three of these chapters consider how teams improvise, sensemake, and cultivate share resources to get their jobs done. These chapters also showcase research that can lead us toward a better understanding of organizing and new media in times of crisis.

Social Media Evolution

Before delving into the first chapter, some context concerning the evolution of social and mobile media is needed. In the 1990s and early 2000s, organizations like corporations and government groups had only a few ways to share information with their publics. Email was common at that time, and so were phone calls and websites. Furthermore, when organizations experienced a crisis or emergency, they had almost exclusive control over how they communicated information and updates *to* their publics: they used news media, direct communication like email, and mass communication such as their own blogs or websites (Stephens & Ford, 2014).

But communication control shifted considerably when the price of mobile devices—such as cell phones—dropped to a level where individuals could afford personal communication devices (Stephens, 2018). That trend, known as enterprise consumerization (Harris, Ives, & Junglas, 2012) combined with the rise in social media platforms, expanded the opportunities to communicate in times of crisis. Citizens could post information publicly, and they could see and engage with information that others shared. Organizations no longer provided updates *to* their publics, but now average citizens could use their new media to share information publicly and self-organize, thus changing how society as a whole was able to respond to emergencies, disasters, and crises (Palen et al., 2010). But the years around 2010 represent the beginning of communication through mobile and social media, and people were trying to figure out how to make them useful, especially in times of crisis. Furthermore, the developers of these platforms were constantly tweaking them based on the needs of their customers and the evolving marketplace (Feenberg, 2009). See Table 0.1 for a timeline of these new media innovations, which I complied by examining company websites, and notice that most of these are relatively new innovations in communication.

This is important context for Chapter 1: the university crises were unfolding at the same time that the U.S. was learning how to use mobile and social media. There were not recognized best practices, and organizations had to constantly adjust to the changing media landscape. Getting a glimpse into what was tried, and refined, is a valuable contribution by Barrett and Posey as we learn from the past and continue responding in times of crisis.

TABLE 0.1 Timeline of social and mobile media adoption

Time period	Event
Late 1990s	Nokia and Motorola offered portable phones
2001	Web 2.0 transitioned from an internal organizational tool to the creation of online communities and user-groups (van Dijik, 2013)
2002–2003	BlackBerry was the first smartphone
2003	Myspace began
2004	Facebook began
2005	YouTube began
2006	Twitter began
	The notion of social came into the Web-based tools conversation
2007	iPhone is introduced
2008	Androids debuted
2010	Instagram and GroupMe are introduced
2012	Facebook reaches one billion users and Snapchat reaches a user base
2013	Twitter goes public
2016–2017	Fake news and misinformation become a major concern on social media

Sensemaking in Teams

Moving into Chapters 2 and 3, the chapters deeply explore ideas of sensemaking under duress. Jahn, a former firefighter herself examines "close calls" and what these firefighters learn from their experiences. She argues that their lived, physical involvements provide opportunities for them to understand their hazardous work. In Chapter 3, Williams shows us what happens when multiple emergency response teams have to work together to save lives. They negotiate for resources, including communication technologies, to accomplish their work.

References

Feenberg, A. (2009). Critical theory of communication technology: Introduction to the special section. *The Information Society, 2,* 77–83. https://doi.org/10.1080/01972240802701536

Harris, J., Ives, B., & Junglas, I. (2012). IT consumerization: When gadgets turn into enterprise IT tools. *MIS Quarterly Executive, 11,* 99–112. http://misqe.org/ojs2/index.php/misqe/article/viewFile/416/313

Palen, L., Anderson, K. M., Mark, G., Martin, J., Sicker, D., Palmer, M., & Grunwald, D. (2010). A vision for technology-mediated support for public participation & assistance in mass emergencies & disasters. In *Proceedings of the 2010 ACM-BCS Visions of Computer Science Conference* (p. 8). British Computer Society, Edinburgh, United Kingdom.

Stephens, K. K. (2018). *Negotiating control: Organizations & mobile communication.* New York, NY: Oxford University Press.

Stephens, K. K., & Ford, J. L. (2014). Crisis communications and sharing message control. In M. Khosrow-Pour (Ed.), *Encyclopedia of Information Science and Tech* (pp. 462–470). New York: IGI Global.

van Dijik, T. A. (2013). News analysis: Case studies of international and national news in the press. Hillsdale, NJ: Lawrence Erlbaum.

Weick, K. E. (1979), *Social psychology of organizing*. Reading, MA: Addison-Wesley.

1

ORGANIZATIONAL CRISIS COMMUNICATION IN THE AGE OF SOCIAL MEDIA

Weaving a Practitioner Perspective into Theoretical Understanding

Ashley K. Barrett and Cindy Posey

In 2010 an active shooter attacked a large public university campus in Texas. A male student, sporting all-black attire and shielded by a ski mask, stormed onto campus with an AK47. As he approached one of the most populated streets of the university, he began shooting. He calmly walked down the street, firing 11 rounds into the ground and sporadically into the air. Not a single student was physically harmed. Afterwards, he ran into the main library on campus, fled up the stairs to the sixth floor, and eventually took his own life.

Throughout this terrifying event, the university campus was on lockdown for four hours. Students and faculty holding class were confined to their classrooms. There, they anxiously awaited further direction. Bus drivers in mid route were informed not to drop students off on campus and others were warned not to approach the premises at all. Students promenading the campus streets heard vigorous instructions to get inside and "shelter-in-place." Family members and other stakeholders at home, at work, and across the globe, were kept abreast of the shooter's whereabouts, the potential second shooter lurking on campus, and other evolving events, by news anchors and television screens.

Messages and notifications of this nature are imperative throughout organizational crises. Serving as vital data points, these messages save lives—but to do so, they must be accurate, timely, complete, reliable, and believable (Pipino, Lee, & Wang, 2002).

It is difficult to imagine that there was once a time when organizations relied primarily on text messages, email, and the phone to communicate such messages to their stakeholders during a crisis. Regardless of our perceptions, this world is not too far removed from our rearview mirror. The 2010 university crisis summarized above serves as one primary example. On this day, which was wracked with uncertainty, the campus police department and university internal communications personnel predominantly used email and text messages to communicate with students, faculty, and staff both on and off campus. As the crisis unfolded, three internal communications operators hunkered down in-between the command post and the media staging area. At this location, they also took calls from the media and wrote additional emails to collect information from other point people on and off campus. In total, faculty, students, and staff received two emails and five text messages throughout this four-hour crisis. It was this crisis communication team's first attempt at using social media during a crisis—five Facebook posts and two tweets—and the university police department's Facebook page went from 469 followers on September 27, 2010 to 10,313 on September 30, 2010.

When the internal communications team debriefed post-crisis, they realized they had each received feedback that people needed to be updated at more regular intervals during a crisis. They concluded in future crises they would send emails to campus members every 10–15 minutes, even if they had not secured any new information. Although they lack new details, "holding statements," as they were termed, can remind people to stay calm and attuned to official campus emergency messages.

In today's densely connected and fast-paced world, the idea of receiving a total of seven official organizational messages (social media posts were not considered official messages in 2010 and with so few followers, they were not very effective) throughout a four-hour crisis would likely unleash a taxonomy of concerns—all having a common denominator of panic. The contemporary university student, or any organizational employee for that matter, is inundated with messages throughout the workday. From iPhones, to Apple watches and Fitbits, organizational members are not only constantly receiving messages, but wearing messages in a manner much more sophisticated than on T-shirts. When messages are this readily available and "sendable," our expectations and desires to receive messages also increases. For example, millennials and other new organizational members increasingly favor more open and frequent communication with their supervisors and enjoy technology use at work (Meyers & Sadaghiani, 2010). Given this growing expectation for communication, receiving the then suggested 10 to 15 emails during a current-day organizational crisis seems hardly enough. What's more, crises are fraught with uncertainty wherein needs for information skyrocket. During organizational crises, members are engaged in a highly active uncertainty reduction process (Reynolds & Seeger, 2005). Previous research has demonstrated that organizational members rarely, if ever, encounter information overload during these equivocal and time-sensitive emergency events (Sorensen, 2000; Stephens, Barrett, & Mahometa, 2013).

In addition to providing mobile access to email and elaborating and expanding the text message game, smartphone use increases the number of messages people send, receive, and seek throughout the day because they serve as mobile gateways to social media sites like Facebook, Twitter, Instagram, YouTube, and Reddit (see Oulasvirta, Rattenbury, Ma, & Raita, 2011). In 2010, the iPhone had only been on the market for three years. The newest iPhone (iPhone 4) included dazzling features that were first being introduced to the public, including email pushes/notifications, FaceTime, front-facing cameras, and advanced non-Apple apps—such as the original Facebook app, which was released in 2010, but was still very buggy and slow (see Casti, 2013). This timeline, coupled with the initial idea that personal technologies should be used for personal reasons, makes it unsurprising that organizations have been slow to incorporate new technologies—like social media—into their crisis communication repertoire. In fact, up until fairly recently, social media has largely been used in an informal capacity as a community information resource during crises. People personally impacted by a crisis often revert to social media to generate and share crisis information in a very timely peer-to-peer fashion (Sutton, Palen, & Shklovski, 2008). This has been labeled crisis backchannel communications (Sutton et al., 2008), and organizations have previously ignored the powerful nature of these online narratives during a crisis due to perceived issues with legitimacy and reliability (Palen,Vieweg, Liu, & Hughes, 2009). However, they are quickly learning that they must not only be aware of these narratives, but also monitor them and integrate them into formal crisis communication if they want to keep their stakeholders safe.

This chapter uses extensive examples and commentary from a crisis communication practitioner, who served as a public information officer (PIO) and in a campus safety communication capacity at a large public university in Texas, to explore the evolution of organizational crisis communication strategies and decision-making in a digital information age stimulated by advanced smartphones and social media. We rely upon the experiences, reflections, knowledge, and predictions of this organizational and crisis communication specialist to visualize the newer, sharper, more erratic boundaries of organizational crisis response. Moreover, we use her purview to more clearly understand the unprecedented, multifaceted communicative challenges impeding organizations from being able to successfully communicate with their publics when they need to most. Despite all of its benefits for information production, collective generation, and accessibility, social and mobile media technology tests and potentially defies practitioners' ability to feel adequately trained or "prepared" for a crisis even with the time, money, and methodical strategizing that are routinely invested into organization's crisis preparation. We use Weick's Model of Organizing (1979) to illustrate how social media use increases human agency and equivocality in crisis communication, thus demanding that practitioners focus on the process of organizing—rather than the static organization—to effectively communicate during a crisis.

Crisis communication practitioners arduously work around the clock to send vital messages when a crisis strikes. Still, the era of social media and advanced personal devices risks increases levels of practitioner burnout and exhaustion during crisis preparation, crisis events, and crisis debriefs/resolution. If we can use theory to understand and predict how new technologies are shifting the nature and necessities of organizational crisis response, perhaps (a) we can decrease the debilitating uncertainty triggering practitioner burnout and (b) organizations can productively and proactively use these interactive channels in a crisis. In the next section, we start by further unraveling how new media have altered organizations' information environment, providing both affordances and constraints to crisis response.

Organizational Crisis Response during the Era of Smartphones and Social Media

Organizations must constantly foster a culture of change to be successful, and technology is often a core component driving and enabling this change (Weick, 2007; Bharadwaj, 2000). Yet, technologies are regularly undergoing change as well, which can create an endless and trepidatious game of catch-up for organizations. Just as organizational leaders feel they have their feet solidly planted underneath them because a technology has been successfully implemented, used, or integrated, the vendor releases massive updates, or a completely new and more sophisticated technology hits the market. Yet, in order to survive and thrive, organizations must master *requisite variety*—meaning their internal operations (policies, processes, routines, technologies, etc.) must be as sophisticated as the external challenges organizations encounter (Weick, 2007). If organizations understand globalized markets and the evolving and expansive nature of technology has complicated issues, they must complicate the nature of how work gets done in their workplace—and which technologies they choose to use. Taking from Karl Weick (1979), organizations must achieve *negative entropy* to remain relevant and to remain in order—they must continue to innovate, to adapt, and to extensively, unconditionally, and wisely criticize and reevaluate workplace operation and communication procedures.

Given that organizations should resemble open systems that accept and adapt to feedback (Weick, 1979), workplace communication is often in flux as policies are reconsidered and renegotiated according to external factors. Smartphones and social media are two primary examples of complicated changes external to organizations, which have impacted the global market, altered societal communicative norms, and thus demanded that organizations reassess and further complicate their internal dynamics. For example, the rapid yet profound evolution in how people send and receive personal and workplace messages has some clear consequences for productivity, health, and the ability to demarcate boundaries. Smartphone use, especially for younger generations, has been likened to addiction (Lepp, Li, Barkley, & Salehi-Esfahani, 2015; Samaha & Hawi, 2016), and linked to high levels of psychological and emotional stress, decreased work performance, and work–home interference (Derks & Bakker, 2014; Samaha & Hawi, 2016). Therefore, organizations must be aware of these new technology pitfalls and create safety parameters for employees' technology use rather than using these devices as controlling and manipulative mechanisms that can eventually elicit burnout (see Stephens, 2018). Yet a common characteristic of any technology is its paradoxical outcomes. For example, although smartphones are harming psychological and emotional health, healthcare organizations are using these devices to simultaneously revolutionize healthcare. To illustrate, one day soon you will be able to use smartphones to help facilitate basic lab tests, thereby acting as a "doctor in a pocket" (Baggaley, 2017). Hence, some organizations understand the potential value of new technologies and cleverly integrating them into routines in innovative ways to enhance products and services and satisfy customers' perpetual needs for versatility, ease, speed, instant accessibility, and frugality. Regardless of their positive or negative consequences, smartphones and app culture have unquestionably and exponentially expanded the number, range, and sophistication of messages that organizational members encounter at work and home each day, and organizations must accommodate this change (see Gao, Liu, Guo, & Li, 2018). However, rather than reacting to the sharply evolving nature in which customers communicate, the most successful organizations will proactively research how to use the smartphone industry to their benefit. Thereafter, they can use these technologies to reach unprecedented types and levels of organizational success.

Successful organizations' modus operandi for requisite variety and drive to avoid entropy (i.e., disorder) can be further illustrated by their response to the introduction and widespread use of all-encompassing social media technologies. Some research suggests these technologies physically impact brain development—causing brains to change and grow in different ways—and affect how the brain works, like how people socially learn, feel rewarded, express themselves, and/or how they're influenced (East, 2016). Given these changes and the impact they could have on organizations' employees, social media use should also impact age-old organizational communication practices, like crisis communication plans.

How It Was and What It Became

Let us for a minute revisit the 2010 crisis example previously explored in this chapter. As you'll remember, seven messages were distributed to organizational employees and students throughout the four-hour lockdown. Internal communication team members received consistent feedback from faculty, students, and staff that they needed more updates to understand how to proceed and to feel securely informed.

Now, fast forward to 2016 when, on this same college campus, a young man not affiliated with the university followed a female student, who was walking to her dormitory one night, and then sexually assaulted, strangled, and killed the student on campus premises. The female student's body was not found until two days after she had been reported missing. The suspect was arrested a few days after the female student's body had been found, and he was charged with capital murder. During this more recent crisis, the university utilized social media sources to send important, timely messages to students, and to collect information on student behavior, emotions, and actions. Shortly after the news of the student's death broke, the university's internal communications department closely monitored several social media sites including Facebook and Twitter. As they navigated these sites, they observed the deep sense of shock and sadness felt by the university's publics, in addition to irrational secondary behaviors and incidents. Due to the observed and warranted fear-filled reactions (during this time, the suspect had not yet been apprehended), the internal communications group and campus police department decided to greatly increase police presence across campus including additional police officers on foot, in cars, on bikes, and even on horses. The internal communications team then turned back to social media waves to monitor the campus' response to the increased police presence. There, they saw a flood of comments describing how safe the police surplus made students, faculty, and staff feel on campus. Unfortunately, social media messages also indicated that some students were not taking their safety seriously. University officials used social media to discover female students who were still walking around campus alone at 2 and 3 o'clock in the morning just one day after the murder.

Social media also played a vital role in the university's post-crisis phase. During this time, university officials launched a campus safety campaign, urging students to walk together across campus and put down their phones to be aware of their surroundings. These campaigns were launched across social media—these platforms were the primary channel through which students learned of the campaign.

As compared to the 2010 crisis, the 2016 example demonstrates what we here coin a *new era crisis*. By 2016, social media use had impacted how people communicate, socially learn, and are influenced, and the university organization constructively and thoughtfully adapted to effectively reach members during a crisis, keeping them safe and informed. The campus police and internal communications staff used these sites to send information to students in the post-crisis phase, but perhaps more insightfully, they used the sites throughout the crisis as powerful and timely *sensemaking* tools. Officials mobilized the social networks to reduce students' crisis-related uncertainty, but also to actively reduce their own uncertainty of how the impacted publics were reacting to and behaving in relation to the campus crisis.

According to Weick, Sutcliffe, and Obstfeld (2005), sensemaking involves the process of understanding a situation through words, and it serves as a springboard for action. People explicitly engage in sensemaking when the current state of their perceived world does not parallel the expected state. In these circumstances, individuals sense the flow of action is unintelligible, thus jolting them out of a state of immersion and flow. To recalibrate, people search for information, interpretations, and reasons to resume activities and get "back on track." The 2016 campus attack and response reveal how official organizational crisis responders used social media to further inform their own interpretations of the crisis and to simultaneously influence students' crisis-related interpretations and actions. Ultimately, these sites were used by the internal communications team to try and "get back on track."

However, as stated before, a common characteristic of new media are their paradoxical outcomes. Although social media and smartphones can enhance certain dimensions of information/communication quality during a crisis, and therefore act as resourceful channels of communication for organizational officials, they can also damage and threaten other dimensions. The 2016 crisis outlined above exemplifies how smartphones and social media avenues enriched and further enabled information/communication accessibility, interpretability, timeliness, and amount appropriateness (see Pipino et al., 2002 for a complete list of the information quality metrics). However, these new, interactive technologies—which provide immediate and highly unregulated authorship and mechanisms for quickly combining and sharing previous messages—can produce egregious errors for information completeness, believability, objectivity, consistent and concise representation, understandability, and value added— all of which are vital in crisis response.

Evolving Social Media Use during Crises

Let's turn to a different crisis communication story to more fully illustrate the complexity in crisis-related social media use. Unfortunately, one year later in 2017, the university encountered another on-campus crisis that ended in a student death. This time a male student suffering from depression walked onto campus with a large hunting knife and went on a stabbing spree, injuring three male students and killing one other. When city and campus police officers were notified, they arrived on the scene *two minutes later*. Once there, they saw the suspect walking away from a stabbed victim, ordered him to the ground, and took him into custody.

This crisis was different from both the 2010 and 2016 crises because it happened in the middle of the school day, and it happened in the evolving age of social media. Hundreds of students witnessed the ghastly crime first-hand and immediately got out their smartphones and opened a social media app to document what they had personally seen and experienced. The initial panic stimulated by the crisis events unfortunately multiplied as a result of different waves of social media use. First, campus police were busy trying to keep people in the area safe and they did not send a text message or social media message to faculty, staff, and students for 20 minutes after the stabbings. At the time, the organization's crisis communication protocol prohibited the internal communications team from publishing information on their social media sites until it was officially released by the police department. During those protracted 20 minutes, multiple administrative communicators attempted to reach the police department, but did so in vain. Police officials were working diligently at the scene of the accident to calm a rattled, chaotic crowd and found it difficult to follow the communication protocol.

The assistant police chief, normally tasked with communicating with the university internal communications team, was at a physician's appointment during the time of the emergency. Moreover, the regular point person on the communications team was on vacation. When crises strike, organizations must communicate with their publics rapidly to not only reduce crisis-related uncertainty, but to establish empathy and reassurance, reduce emotional turmoil, designate crisis spokespeople, offer a general, broad-based understanding of crisis events, and to broadcast specific information on emergency management and medical community responses (Reynolds & Seeger, 2005). On any other day, 20 minutes may not feel like an extended period of time, but during a crisis and in the age of social media, it can seem like an eternity. Without an official organizational response, students were left to seek information from informal channels (e.g., friends' social media sites) on their own accord.

Crisis communication is often plagued with quality issues because crises occur in a context of skyrocketed uncertainty wherein each minute or even second that passes is crucial. However, a lack of formal organizational communication immediately following the initial crisis event can significantly escalate these predisposed issues, leading to additional problems with information completeness, consistency, objectivity, and believability. Hence, it is not surprising that when there were problems with the communication protocol, the internal communications team's ability to use social media as tools for information distribution was stalled.

In addition to the slow formal organizational response, panic during the crisis' initial stages was exacerbated as students turned to social media networks to propagate and rapidly circulate unconfirmed crisis details. For example, the campus fraternity community had recently undergone intermittent periods of scrutiny as it had been identified as contributing to the college rape culture epidemic frequently discussed in the popular press at that time. Fraternity buildings and property had been vandalized in previous weeks. Some fraternity members had even received threats of violence. Immediately following the stabbing event, social media and GroupMe apps were flooded with accusatory messages that the stabber had intentionally targeted students in the Greek community. Fraternity and sorority members on campus who wore their Greek letters across their chests and backs frantically ran into buildings to turn their shirts inside out. Student government officers were breathlessly warning fellow Greek students to run home to their dorms or apartment buildings and hide. Parents began calling radio stations and administrative phone numbers in droves. The Greek ties to the stabbings turned out to be completely fabricated and based on suspicions and conjectures.

As the false messages about the attack's Greek roots were spreading like wildfire across social media and GroupMe, another mentally unstable student faked a stabbing incident in a neighborhood filled with student apartments close to campus. Yes, this happened shortly after the other stabbing and the unfolding events. This student stabbed himself in the leg, and afterward he turned to social media to announce that he was a hero: he had intervened in a stabbing of a sorority girl, saving her life. Messages were swirling on multiple social media platforms, but this was also a fabricated event.

Finally, co-occurring with these multiple events, a battle was also brewing on campus between radical leftist and radical right-wing groups. The members of these groups claimed to be students, but university officials were aware that the group members were not affiliated with the university; rather they were outsiders trying to gain a foothold on campus and make social capitol strides. Allegedly, a radical leftist group member had illegally hung a banner on a campus pedestrian bridge that morning. The language on the banner was mildly threatening. Sometime after the banner was hung, someone ran through a university building adjacent to the pedestrian bridge yelling, "Bomb threat. Evacuate now!" News of the banner and yelling contagiously diffused through social media sites, creating tributary crisis narratives that exacerbated the larger stabbing crisis story already forming online. Unfortunately, social media posts from multiple individuals on campus stirred even more unnerving confusion in regards to the relationship between the stabbing and these hypothetical bomb threats.

As evidenced above, multiple incidents occurring alongside the 2017 stabbing crisis—all involving the use/misuse of social media—made this day a communications nightmare for the university. Perhaps the number and pervasive nature of the fabricated stories flooding social media networks were a direct result of the university's delayed crisis response, but it is difficult to know because misinformation can spread even when responses are timely. Previous scholarship urges organizations to "partner with the public" in a crisis, meaning organizations should deliberately share crisis information in a timely and honest manner to (a) ease uncertainty and (b) meet the publics' communication needs so they do not turn to other sources (Veil, Buehner, & Palenchar, 2011, p. 211). If an organization is silent or is deemed by its publics as an untrustworthy source of information for any other reason, its ability to manage the crisis is fleeting (Veil et al., 2011). The immediate, unregulated nature of social media implies that, if formal crisis information is not quickly distributed by organizations, the backchannel crisis narratives formed by informal sources online will gain more predominance; yet these narratives can be emotionally fueled, not fact-checked, and therefore illegitimate (Sutton et al., 2008). Thus, there are pros and cons to crisis storytelling using social media. In the section below, we discuss how social media has specifically impacted the composition of storytelling, which further emphasizes how social media sites can be both advantageous and precarious communicative tools for organizational crisis response.

Social Media and the Transformed Nature of Crisis Storytelllny

Storytelling has traditionally been defined as orally presenting causally related events or an experience in temporal order (Hughes, McGillivray, & Schmidek, 1997). Some scholars have gone so far to label stories the "sensemaking currency of organizations" (see Rosile, Boje, Carlon, Downs, & Saylors, 2013). Yet stories told through the waves of social media rely upon nuanced features that expand previous traditional notions of storytelling. To illustrate this point, we discuss three of these features specifically, labeling them: *distributional authorship*, *digital recombination*, and *reviewability*. In the next few paragraphs, we define each of these storytelling characteristics and explore how they might manifest in a crisis storytelling event.

First, *distributional authorship*, a term coined here but implicit in the digital storytelling literature, refers to the idea that as technology is changing the boundaries of organizations, it is also renegotiating the authorship boundaries of organizational stories. Initial theoretical—and very influential—work on narratives claimed stories should be well structured, have a beginning, middle, and end, and are monologic—including one (or very few) clear narrator(s) and a listening audience (Labov, 1972). Conversation analysis researchers will claim story construction has always had a degree of "co-tellership" (De Fina & Georgakopoulou, 2012, p. 45). Yet, the expansive use of social media networks has further discredited the age-old idealization of a story having a single, or even a few, author(s).

Social media features are designed to support diverse participation frameworks, the co-constructing of emergent stories, and stories that are told *in situ* through collective interaction (see De Fina & Georgakopoulou, 2012). Different technologies provide different contexts (i.e., perceived affordances) for storytelling that directly impact the structure of stories (Georgakopoulou, 2017). With social media technologies, story creation largely becomes a communal authorship project as individuals in the same or different places and times contribute their story details from the comforts of their own personal computers. Moreover, social media have expanded the boundaries of story authorship in that they afford more voices to be heard in a short amount of time—even simultaneously. Social media's affordances—for example, instant status updates, audio-video posts, groups with visible discussion threads—enable participants to rapidly shift from audience to author (and back) in story creation. Authorship is highly unregulated in these interactive social media platforms—all it takes is the click of a button and story details are publicly visible. Consequently, social media sites become home to a diverse set of published perspectives (see Sobkowicz, Kaschesky, & Bouchard, 2012).

Social media's property of collective, speedy story building, along with democratized publishing, can be an asset for organizations and their stakeholders during a crisis. For one, it enhances employees' perceived self-efficacy—or their perceptions that they are capable to perform tasks that matter or help others during a crisis. Increasing stakeholders' self-efficacy has been categorized as a best practice in organizational crisis communication (Veil et al., 2011). History consistently attests to the fact that impacted populations practice resilience following a tragedy by actively participating in response and recovery efforts. Highly accessible social media platforms designed to easily facilitate information creation and sharing furthers the public's capacity to engage in organization's response and recover efforts (Palen et al., 2009). For example, Palen and colleagues (2009) discovered that after the Virginia Tech shooting massacre in 2007, students engaged in wide-scale problem-solving efforts online wherein they rapidly converged information to build a list of the deceased. Moreover, this crisis story (i.e., a peer-generated list) was more accurate and timely than those generated and authored by the popular press. However, distributed authorship can also have negative implications for organizations. As observed in the 2017 university stabbing crisis discussed above, social media's affordance of democratized, unregulated publishing can convert into the rapid dissemination of unconfirmed and/or false story details during a crisis. These fabricated stories—i.e., the example of the mentally unstable student publishing a story on social media claiming he had been stabbed in the process of saving a sorority girl's life—create noise that deter from the organization's ability to effectively communicate crucial crisis information to its publics. This particular disadvantage can be crippling for organizations, given that combatting noise during a crisis was already a daunting challenge before the age of social media (see Stephens et al., 2013).

Second, and innately linked to distributional authorship, social media afford *digital recombination* in crisis storytelling. In delineating the integrated structure of storytelling, Bruner contends:

> an individual's working intelligence is never "solo." It cannot be understood without taking into account his or her reference books, notes, computer programs, and data bases, or most important of all, the network of friends, colleagues, or mentors on whom one leans for help and advice.
> *(Bruner, 1991, p. 3)*

As this quote contests, the notion of *recombination* in story creation preceded the digital era. Rather it expands on previous narrative notions of *entextualization*. Entextualization consists of "lifting a piece of discourse out of its context of production and inserting it into new contexts in which it receives a new interpretation" (De Fina & Georgakopoulou, 2012, p. 133). Thus, through this process of entextualization, formulated stories are reformulated, uplifted, and implanted into new contexts wherein they'll receive a new, or advanced, interpretation.

This process of entextualization is aggrandized and refined in story creation on social media, wherein tellership is socially constructed and authorship is decentralized and unregulated. Social media provide a hyperactive atmosphere for reformulating stories and for engaging in *online* entextualization—hereafter labeled *digital recombination*. These highly trafficked story-generating venues are comprised of an infinite number of communicators taking previous updates, statuses, knowledge (i.e., formulated stories), adding their own personal contributions to these stories, and then publishing them for others to thereafter interpret (Faraj, Jarvenpaa, & Majchrzak, 2011). For example, in a crisis situation, organizations often create private Facebook groups to provide impacted populations with a virtual location to engage in information sharing and updates (see White, 2012). These groups contain a "wall" where group participants can post evolving information about the crisis. Digital recombination would occur when someone shares a popular press story on the group's wall and then adds a personal message summarizing the most important parts of the article or offering an emotional interpretation of article content.

The above example represents just one case of digital recombination. Contributors can also engage in *story-linking* (Georgakopoulou, 2017) wherein they contribute to previously formulated stories by hyperlinking material into previous status updates or group conversations. Organizations can also hyperlink or share stories onto their official social media pages to spread helpful information from community stakeholders like police departments, or other federal agencies involved in the crisis response. Moreover, organizations can share fabricated stories on their page, including a clear additional message that emphasizes the story's incorrect details. This fact-checking example leads to our next and final example of how social media affordance change crisis storytelling—reviewability.

Reviewability is a social media affordance referring to these platforms' ability to enact knowledge-collaboration forums in which active participants are able to monitor, review, and manage the collaborative content of emerging crisis stories. Organizational stakeholders participating on social media sites can embrace "role-making" behaviors (Goffman, 1959) in which they adopt a self-assigned responsibility such as organizer, supporter, editor, or corrector. Therefore, reviewability can improve a story's verisimilitude, or the likelihood that the story's anatomy and collective meaning mirror reality (Bruner, 1991). This is crucial during times of crisis or high uncertainty wherein credible information is vital. Crisis storytelling on far-reaching social media sites implies that the number of active audience members scanning, surveying, and auditing emerging stories skyrockets, and consequently, so does reviewability (see Hutchby, 2014).

Moreover, as previously mentioned, organizations can proactively scan, monitor, and correct information published by both internal and external populations during a crisis. In fact, after the devastating rumor mill effects that impeded the organization's ability to effectively communicate with it publics in the 2017 stabbing crisis, the university's internal communications team decided to launch an information credibility campaign. This campaign, spread by the university's official social media sites, included materials that were designed to educate faculty, students, and staff how to properly evaluate information on social media sites following a crisis. These materials urged organizational members (a) to remember first reports are often incomplete and inaccurate, (b) to not trust third-party sources, (c) to be aware of image manipulators, trolls, and other information fakers, and (d) to understand that language like "we are getting reports that…" means that this information has not yet been confirmed.

Acting as an offshoot to reviewability, crisis storytelling on social media also affords *transparency*. Social media technologies include features that facilitate the presentation, storage, retrieval, and flow of information in ways that is difficult, if not impossible, to accomplish face-to-face (Leonardi & Vaast, 2017). Social media platforms represented multiple user-generated information environments, wherein multiple participants not only monitor and edit story details, but these details and their updates are visible to the public, sometimes indefinitely. In addition to this *communication transparency*, social media environments also afford *identity transparency*—or the ability to observe a communicator's profile information, demographics, liked groups, historical data, etc. (Stuart, Dabbish, Kiesler, Kinnaird, & Kang, 2012). For example, in the Facebook group instance offered above, one participant might post updated information about the names of crisis victims to the group wall. Yet, depending on privacy settings, sometimes these groups can be plagued with "trolls" who post fake information to the group's wall or distract crisis story building with irrelevant political rants. Because the name of crisis victims is a crucial element to the crisis story, other participants may choose to confirm this new participant's identity. They can do so by clicking on the new participant's homepage and determining and searching for important details such as how old they are, where they go to school, where they work, previous posted images, Facebook profile pictures, etc. Identity transparency can of course be limited due to privacy settings. Even so, the sources behind information posted on social media sites are often more visible than those contributing to popular press stories.

This section of the chapter exemplified how crisis storytelling in the age of social media poses several affordances, but also constraints, for effective organizational crisis communication. Distributed authorship, digital recombination, and reviewability represent three examples of how crisis story creation and evaluation has shifted, and organizations must not only adapt, but learn how to take advantage of this shift. As evidenced, organizations can use these sites to quickly disseminate information in a one-to-many fashion following a crisis event. Yet, organizational officials must also acknowledge the ubiquitous and inherently interactive design of these technologies, viewing them as many-to-many platforms that collectively and dynamically create and share information during a crisis. Thus, social media use forces organizations to conceptualize crisis communication plans as in flux; they evolve with the unique nature of each crisis and the unique features and reactions of impacted populations. Universal, static crisis communication plans and protocols are quickly becoming a concept of yesteryear. In the age of social media, unregulated authorship, and rapid online information convergence, contingencies in organizational crisis communication strategies are to be expected. With these advantages and challenges in mind, the next section explores how Weick's Model of Organizing can be applied to the stages of a crisis and reveals how a focus on organizing is critical to effective crisis communication in the era of omnipresent social networks.

Using Theory to Understand the Evolution of Crisis Response in the Social Media Age: A Focus on Organizing

In his book, *The Social Psychology of Organizing*, Karl Weick (1979) argues that organizations do not statically exist as something that we can tangibly touch, prescribe consistent behaviors to, or predict with a high degree of certainty. Rather than existing, organizations are always in flux; they are in the process of existing. Organizational communication and human agency are the primary vehicles that drive this ongoing state of organizing. To construct his model, Weick jointly relies upon sociocultural evolutionary theory, information theory, and systems theory—interpreting each of these theories with an organizational lens (see Kreps, 2009). In the next paragraphs, we describe Weick's model in detail before explaining how the framework is directly applicable to unfolding crisis situations. Particularly, his model provides insight into the complicated role social media play in organizational crisis preparation and response.

Weick's Model of Organizing

To advocate for a focus on organizing (i.e., processes), rather than organizations (i.e., structure), Weick (1979) first asserts that organizations are in a constant state of change. Whether that change be internal or external, social or perfunctory, planned or unplanned, organizations must adapt to these changes in their environment in order to survive. These adaptations can manifest as psychological or behavioral, displaying as new organizational strategies or innovative cognitive orientations, for example. However, organizations encounter more than unremitting phases of change and adaptation; they are also in a perpetual state of learning and retaining, establishing and growing an organizational memory that will impact future actions and organizational decisions. This organizational memory functions as a series of causal maps that will increase organizational certainty during similar forthcoming organizational changes.

Figure 1.1 visually depicts Weick's (1979) Model of Organizing, and the relationships among its critical stages. As it demonstrates, when organizations experience a change in their environment, they enter into what Weick describes as the enactment phase. In this phase, organizational employees are engaging in sensemaking to reduce uncertainty and organize the organizational flux (Weick et al., 2005). Organizational actors are trying to understand the dynamics of the change, and most specifically, how it will directly impact them and broadly impact the organization. Yet, it is important to emphasize that Weick portrays organizational actors as *enacting*—rather than reacting to—change. To clarify, organizational employees have a primary, active role in constituting organizational change because change is socially constructed—its existence depends on human agency/action, and human communication. Change sensemaking and meaning-making is a product of jointly constructed understandings of a change, which form change realities. These realities form the basis of human action and rationalization, yet, multiple (conflicting) realities can exist. If an organizational change is complex (i.e., if information inputs are ambiguous), it defies actors' ability to understand the

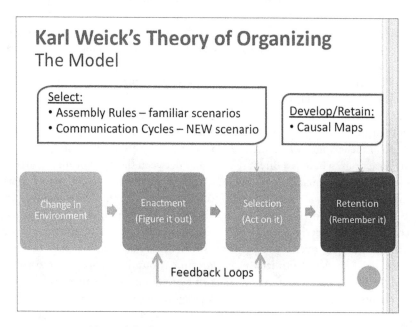

FIGURE 1.1 Weick's Model of Organizing

change; hence, multiple interpretations of the change will likely materialize. Weick (1979) defines this state as the organizational environment experiencing *high equivocality* (or high uncertainty). If there are multiple ways to make sense of a change, organizational actors must engage in vigorous information processing before they can choose a specific change reality, move forward, and ultimately to adapt (see selection stage below). Conclusively, the enactment phase consists of organizational actors engaging in information processing (i.e., sensemaking) in an attempt to reduce equivocality in an organization's information environment.

Change enactment should eventually and organically lead into the selection stage wherein organizational members select one change interpretation and behave and cognitively rationalize accordingly. This process of narrowing echoes the process of organizing (i.e., information processing) and is only possible through emergent human communication. In this stage, organizational employees select a change reality, thereby significantly easing the equivocality in the organizational information environment and de-escalating sensemaking processes. Actors decide what the change information means for them and thereafter commence in the adaptation process. To select a change interpretation, actors can rely upon assembly rules and/or communication cycles. Assembly rules represent pre-set responses and are used when initial states of equivocality in the information environment are more modest. When organizational actors depend on assembly rules to make sense of a crisis situation, they are essentially accessing standardized procedures for familiar situations the organization has experienced in the past. These somewhat automatic and stock responses are derived from the organization's working memory and can quickly enable organizational actors to make sense of change and proceed forward.

Conversely, organizational actors engage in communication cycles when a change wracks the organizational environment with uncertainty. If crisis events—or offshoots of these events—are unprecedented, organizational actors will earnestly engage in information processing—introducing ways to make sense of the change, for example, and reacting to others' change interpretations and reactions. There is a direct relationship between the equivocality in an organization's information environment, and organizational actors' use of communication cycles to respond to information inputs (Kreps, 2009). However, Weick (1979) theorized that changes that were once ambiguous and required communication cycles should become routinized and less equivocal if they occur repeatedly over time. For him, successful organizing demands striking a balance between routinized procedures (i.e., routinizing actions into rules) and flexibility (i.e., retaining a degree of equivocality in organizational environments). This process of transforming communication cycles into assembly rules represent characterizes the final stage in Weick's (1979) model, labeled retention.

In the retention stage, organizational actors reflect on their previous response(s) to change/information inputs and decide whether these response patterns should be repeated if another, similar change tackles the organization's environment in the future. Organizational actors' actions, behaviors, and cognitive frameworks in the enactment and selection stages are gathered, evaluated, and stored, thereby adding to the collective organizational memory. If communication cycles effectively reduced uncertainty and enabled employees to successfully navigate organizational change, they become assembly rules. Thus, a repertoire of rules continuously emerges, composing organizational intelligence. This organizational intelligence takes the form of employees' causal maps that guide future organizational and individual actions. Feedback loops connecting the enactment and selection stages to the retention stage (see Figure 1.1) allow organizational members to maintain a homeostatic balance in organizing processes (Kreps, 2009). Essentially, the retention phase (i.e., the organization's ever-growing intelligence) is used to guide activities and coordination in future enactment and selection stages. In this way, feedback loops are the mechanisms through which communication cycles become rules, enabling organizations to locate an equilibrium between standardized processes and adaptability.

Using Weick to Understand Current Challenges in Organizational Crisis Communication

As previous scholarship suggests, Weick's (1979) Model of Organizing acutely applies to organizations' crisis management and the enacted sensemaking and prominent communication practices found therein (Maitlis & Sonenshein, 2010; Weick, 1988). As earlier noted, sensemaking is triggered by disruptions in an organization's ongoing environment, or high states of equivocality. Although a variety of organizational changes can solicit high degrees of equivocality (i.e., ambiguity, disorientation, confusion, emotional reactions), organizational crises often carry the highest levels of equivocality and hyperactive sensemaking needs (Maitlis & Sonenshein, 2010). After a crisis—or an unexpected change to the organization's internal or external environment—occurs, Weick's description of the enactment, selection, and retention congruently reflects the actions and processes vital in the three phrases of crisis management—(1) pre-crisis, (2) crisis response, and (3) post-crisis. Specifically, the enactment and selection stages mirror decision-making in crisis response, and retention exemplifies the feedback loops inherent between the post-crisis and pre-crisis phases. Weick's model can also be used to help us understand the complexities to organizing introduced by social media technologies, and organizations successfully communicating with their publics following a crisis. In the next two paragraphs, we (a) elaborate on these comparisons, (b) further detail the content of the crisis management phases, and (c) discuss the challenges social media poses to organizational communication within these phases.

Crisis Response Phase: Elevated Sensemaking

Put simply, this phase encompasses what organizational officials do, and how they communicate with organizational stakeholders after a crisis has hit. Previous research has divided this phase into two pillars—how organizations respond immediately after the trigger event and how they work to assuage reputational damage and keep the public informed in the days, months, or even years to come (Reynolds & Seeger, 2005). After the initial event, organizations are encouraged to rapidly communicate with the impacted publics to reduce crisis-related uncertainty, provide assurance, and increase the public's self-efficacy. For instance, during the initial response, organizations should: (a) designate specific spokespeople and communication channels; (b) provide understandings of crisis circumstances, events, and anticipated outcomes; (c) increase their publics' self-efficacy by telling them what they can do to get involved or get more information; and (d) provide their publics feedback and corrections on any misunderstandings or rumors, to name a few (Reynolds & Seeger, 2005). Looking past those initial moments or hours, the crisis response phase also entails organizations devising a specific communication campaign directed toward affected groups and the general public. Among other things, this campaign should inform the public of the selected recovery, remediation, and rectification efforts and persuade the public they will indeed be successful.

The crisis response phase echoes the enactment and selection phase for several reasons. First, organizations are involved in hyperactive sensemaking after a crisis hits. They are responsible for how that crisis is enacted both internally to organizational members and externally to impacted publics and the general public. Moreover, during the crisis response phase, organizational officials are catapulted into a binary stage of sensemaking. They must learn how to make sense of the crisis themselves, and also have their pulse on the publics' needs for sensemaking. As discussed, change sensemaking and meaning-making is a product of socially constructed change realities, which become the foundation for human action and rationalization. Yet, multiple (conflicting) realities can exist. Organizational crises are undoubtedly complex, creating highly equivocal environments with multiple potential realities. Therefore, it is crucial that organizations work to design these realities by engaging in communication with their publics and representing trustworthy sources of information that provide assurance and reduce emotional turmoil. Organizations must also be competent and strategic communicators to break through the initial noise surrounding the trigger event (Stephens et al., 2013).

Moreover, the crisis response phase mirrors Weick's selection phase because organizational officials must ultimately decide, or select, specific interpretations of the crisis for themselves, their employees, and the public. After a crisis strikes, organizational officials use assembly rules and communication cycles to first choose their specific recovery, remediation, and rectification efforts, and second, to inform their publics of these plans through intricate communication campaigns. Because no two crises are alike, it is likely that organizations will rely upon communication cycles to determine the minute details of crisis response strategies and to effectively communicate these strategies to specific audiences. However, organizations can also rely upon assembly rules in the form of previous crisis communication protocols (the structure of crisis management teams, processes for designating a spokesperson, templates for crisis messages, lists of mandated actions following a crisis, etc.) as catalysts for organizational action. It is also easy to imagine how documented errors in previous assembly rules and/or communication cycles successful in reducing uncertainty can thereafter be (re)formulated into rules that guide the organizational responses in subsequent crises.

Social Media's Impact on Crisis Sensemaking and Enactment

Weick's (1979) model paired with our previous exploration of how social media impact crisis storytelling provides insight into why crisis sensemaking can be stunted during modern-day crises. Put very simply, social media changes the nature of crisis enactment—both for organizations and for their publics. Distributional authorship, digital recombination, and reviewability represent social media storytelling features that alter how people experience and socially construct a crisis. Not only do they augment the amount of information available during a crisis, aptitudes for unregulated authorship and the instant sharing and hyperlinking of information exponentially increase the probability for rumor dispersion. In the same vein, these features can impede an organization's ability to control crisis narratives and affected population's crisis realities.

Think back to the 2017 stabbing crisis previously delineated. During the immediate crisis aftermath, social media platforms granted organizational actors the opportunity for a publicized voice, which eventuated in rumors about a Greek connection to the stabbing and a disturbed student claiming his self-stabbing was related to the original crisis. Social media's democratized authorship and multiuser generated information can contest an organization's ability to manage the crisis-response phrase. Essentially social media use can generate *crises within a crisis*, multiplying the need for accurate information, but confounding the public's ability to differentiate between credible and deceptive sources. Organizational officials can shrewdly use these venues to reduce crisis-related uncertainty/equivocality through the rapid distribution of necessary crisis instructions and facts and through messages that provide assurance, empathy, and self-efficacy. Moreover, impacted populations can use social media to virtually band together, provide each other emotional support, quickly ask for help, and to find other means of fruitfully participating in the crisis (see Jin,

Liu, & Austin, 2011; Wilensky, 2014). However, informal, illegitimate back narratives constructed on these social sites can increase emotional turmoil, strip affected populations of their efficacy, increase crisis-related uncertainty, and stifle organizations' and their publics' sensemaking (see Endsley, Wu, Reep, Eep, & Reep, 2014; Sutton et al., 2008).

Organizations must be aware of these "back narratives" and their levels of veracity to help keep their members safe and properly notified after a crisis event. As for the 2017 crisis, university officials were largely unaware of the backchannel narratives emerging online until debriefing and data collection (i.e., student, faculty, and parent interviews) in the post-crisis stage. Recent literature suggests that, comparatively, organizational officials are less likely to use social media to communicate with their publics during a crisis than other sources. Specifically, local news media diffused the most frequent and timely crisis-related messages throughout these interactive sites and were the most likely to mine these sites for crisis information immediately after crisis events (Chauhan & Hughes, 2017). This positions local media players as prime sense-makers and sense-givers during periods of heightened public uncertainty and information seeking; it also intimates a lack of organizational control that should concern organizational crisis communicators.

Retention Stage: Connecting Pre- and Post-Crisis Phases

According to Weick (1979), organizational actors evaluate their actions, behaviors, and cognitions in the previous enactment and selection stages, thereby engaging in a process of collecting feedback. They decide whether the communication cycles they engaged in did in fact reduce equivocality in the information environment. If the cycles helped organizational actors to reduce uncertainty and effectively make sense of the change, these cycles are transformed into assembly rules in the retention phase. New rules emerge after each crisis, adding to the organization's intelligence and organizational actor's causal maps. These rules and elaborated maps then become principal tools for the organizations crisis planning and crisis preparation.

Essentially, Weick's (1979) retention stage encapsulated the inherent mutually influential relationship between the post-crisis—or evaluative—phase and the pre-crisis—or preparation—phase. In the post-crisis phase, organizational officials engage in methodical discussions on the adequacy of the crisis response, replete with assessments of organizational communication effectiveness (Reynolds & Seeger, 2005). They work to form a consensus on the lessons learned and to ensure these lessons are documented, formalized, and communicated to internal and external stakeholders (Reynolds & Seeger, 2005). These lessons develop into a list of best practices that prescribe specific actions to improve crisis communication and an organization's response capacities. The post-crisis phase is directly linked to the pre-crisis phase wherein organizational officials (a) collect data or feedback, (b) integrate lessons from this data into response plans, and (c) generate communication and education campaigns that are targeted to specific audiences (Reynolds & Seeger, 2005).

As an example, in debriefing sessions associated with the 2017 stabbing incident, university officials learned that the back-up communicators had not had enough training. The internal communications team immediately implemented communication tabletop exercises and engaged in these preparation sessions on a regular basis. Mock scenarios were played out minute-by-minute with primary, secondary, and tertiary players in every possible communication role at the table. The learning outcomes of these sessions, combined with fact that the university police department had not followed the text-message protocol, provided the three lessons learned:

1. The university's police department must send out a text and social media messages immediately.
2. Back-up communicators must be highly trained and understand how to intercept false information on social media.
3. The protocol inhibiting the internal communications department from communicating any information outside of that first offered by the police department should be abrogated.

Additionally, both the 2016 and 2017 university crises evidenced strong feedback looks from the post- to pre-crisis stages. The 2017 stabbing crisis resulted in a communication campaign directed primarily toward students, educating them how to decipher between credible and suspicious crisis messages propagating social media sites. The 2016 post-crisis phase produced a "Be Safe Campaign" to encourage students to walk in pairs or groups and take stock of their surroundings when walking through secluded areas of campus.

Social Media Use Increases Need for Communication Cycles and Interrupts Feedback Loops

Feedback loops are undoubtedly important mechanisms in Weick's theory and in organizations' crisis management. Weick borrowed his notion of feedback loops from systems theory. Positive and negative feedback between system components is vital for systems to achieve a homeostatic balance and to assist or constrain system processes according to the needs of the entire system (Weick, 1979). In crisis management, seeking feedback from organizational stakeholders and the general public is crucial for organizational learning and for public relation motives. The director of the internal communications team at the university of focus in this chapter claimed everyone has an opinion on how crises are managed—faculty, students, university leadership, parents, staff, the public, cooperating agencies—and they are all represented in the debriefing session. Even if these opinions are uninformed, it is important to hear them, and to make these sessions visible to the public, because perception is reality.

Organizations must find clever and transparent ways to adjust their crisis plans and future responses based on the perceptions of the masses. Moreover, these processes of accumulating feedback and intelligence should not only occur in the post-crisis phase. Organizations should also monitor feedback from its stakeholders as a crisis unfolds to effectively adapt organizational strategies and tactics to the stakeholders' needs (Bernstein, 2016). The 2016 crisis exemplified this point when the university's internal communications team monitored students' social media posts during the crisis and subsequently armed the university grounds with a more robust police force. Increasing the sophistication of organizations' use of evolving social media sites, and teaching organizational actors to perceive the affordances of these new technologies (see Leonardi & Vaast, 2017) are certainly current and ever-approaching obstacles for organizations' crisis communication efforts.

However, an amplified need for training and learning is not the only obstacle social media use engenders for organizational crisis planners and communicators. As discussed earlier, these networking sites intensify human agency during a crisis by enabling just about anyone to contribute to evolving crisis stories, and in doing so, becoming agents in crisis response. This distributed authorship has advantages for timely information convergence (see Sutton et al., 2008) but several daunting challenges—including information legitimacy as explored earlier in this chapter. Yet another unsettling correlative of stakeholders' social media use is that organizations can experience unparalleled discrepancies among crisis events. As the director of the university's internal communication team noted, crises have always had their distinct elements; but in the age of social media, the pathways down which crises and crisis communication can unfold are infinite. Consequently, organizations' abilities to predict the nature of crises and to formulate effective crisis communication plans can be diluted (Sutton et al., 2008). One could convincingly argue that these sites enable organizations to communicate directly with their publics, and thus grant organizations more control over crisis communication than traditional media sources wherein journalists determine message framing. Despite this uninterrupted, speedy access to the public, social media has led to ever-increasing equivocality in crisis response—a process already convoluted with equivocality. This heightened uncertainty nulls the use of previous assembly rules and augments needs to engage in communication cycles, thus threatening the homeostatic balance of organizations. Moreover, feedback loops connecting previous lessons learned to current-day activities can have less of a shelf life, given social media sites are rapidly updated with new features. Thus, an organization's capacity to be and feel prepared can be diminished along with organizational actors' abilities to construct operative causal maps.

The 2017 crisis exemplifies this point—how can an organization prepare for counterfeit narratives of stabbings and bombs that so quickly and automatically

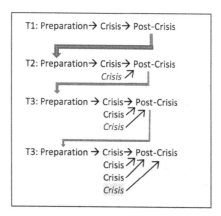

FIGURE.1.2 Social media's impact on an organization's feedback loops

★ Note. Thinner arrows represent less prepared organizational responses. The italicized crisis represents the additional crises for which organizations are utterly not prepared. The multi-angled arrows represent feedback loops from the post-crisis phase of one crisis to the preparation phase of the next crisis. Straight arrows indicate an influential relationship.

reach organizational stakeholders? The university can, of course, create campaigns educating stakeholders on online information credibility, thus integrating feedback into crisis preparation and future enacted sensemaking. However, as social media increases stakeholders' agency in crisis communication, it can impede and deter from formal organizational crisis communication messages. Moreover, as Figure 1.2 indicates, these interactive sites enhance the potential for crises to occur within an existing crisis because they (a) expound the number of crisis messages stakeholders must filter through, and (b) give informal, unchecked information a center stage.

Conclusively, social media can force organizations into a primarily reactive, rather than proactive, state in the crucial minutes or hours shadowing a crisis. They can increase equivocality in the information environment, aggravate an already desperate need for sensemaking, and negatively impact an organization's ability to constructively use feedback loops, thereby transforming communication cycles into future assembly rules.

Implications and Future Directions

This chapter used insights and examples from a university crisis communication practitioner to explore how social media has influenced organizations' crisis communication and response. Given social media can exacerbate equivocality in organizations' crisis response and push organizations into unprecedented reactive states, organizations must adapt, making deliberate and astute strides in preparation.

Practitioners who find themselves in these crisis situations can focus on productive activities that are often generated by crisis communication team's own feedback loops. For example:

1. First, organizational communicators must conduct more tabletops of their own. Emergency preparedness does a great job of conducting tabletops that include everyone—fire, EMS, hospitals, organizational personnel, and the official communicators. However, those drills tend to focus on the first responders, as they should. Yet an organization's official crisis communication team should routinely perform communicator tabletops to play out various bizarre scenarios and mimic and ascertain the feel of the crisis moment. It is important to play out the crisis scenarios minute-by-minute to discuss options regarding how crises should be handled. Social media exercises and potentialities should be discussed along with an exploration of how these sites are currently being used by organizational stakeholders. Participants have left the tabletop emotionally spent, as if the crisis communication scenarios had actually occurred in real life.

2. Second, although it is imperative that organization's crisis communicators receive enhanced social media training, the grueling nature of new-era crises make it more important than ever that they also receive mental health crisis training and access to mental health treatment post-crisis. Although first responders often receive this training, no programs currently and specifically concentrate on organizational crisis communicators. This is problematic given that social media have only intensified the scrutiny that communicators are under during a crisis. These social, democratized authorship venues offer an easily accessible and visible venue for parents, the students, the faculty, staff and other organizational stakeholders to constantly comment and criticize official communicators' work.

Looking forward, future research should continue to thoughtfully unravel how new, ever-evolving technologies like social media and smartphones provide constraints and affordances to organization's crisis communication. Moreover, how do these technologies reinvent older notions of crisis response? For example, the university communications team highlighted in this chapter collected anecdotal evidence that students do not dial 911 during a crisis. They rather use text messages and social media to distribute crisis messages. In fact, many smartphone companies are now providing the public with an option to text 911. Future research should examine the implications of younger generations' potential inhibitions with 911 for organizations that must deliver timely, accurate information during a campus crisis.

Discussion Questions

1. Which social media sites are best and least equipped to aid in an organization's crisis communication response based on their features and reach?
2. What specific protocols and practices can organizations employ to ensure their crisis communication strategies continue to evolve with the production and reproduction of social media channels?
3. What are some advantages and disadvantages of organizations not having full control over how their internal and external publics respond to an organizational crisis?
4. How can crisis communication teams help organizational stakeholders decipher between credible and counterfeit crisis messages both on- and offline following a crisis?

References

Baggaley, K. (2017, October 13). Smartphones are changing medical care in surprising ways. *NBCNews.com*. Retrieved from: www.nbcnews.com/mach/science/smartphones-are-changing-medical-care-some-surprising-ways-ncna810561

Bernstein, J. (2016). The 10 steps of crisis communication. *Bernsteincrisismanagement.com* Retrieved from: www.bernsteincrisismanagement.com/the-10-steps-of-crisis-communications/

Bharadwaj, A. S. (2000). A resource-based perspective on information technology capability and firm performance: An empirical investigation. *MIS Quarterly, 24*, 169–196. Retrieved from: www.jstor.org/stable/3250983

Bruner, J. (1991). The narrative construction of reality. *Critical Inquiry, 18*, 1–21.

Casti, T. (2013, August 1). The evolution of Facebook Mobile. *Mashable.com*. Retrieved from: https://mashable.com/2013/08/01/facebook-mobile-evolution/#SXbEF5guj8qi

Chauhan, A., & Hughes, A. L. (2017). Providing online crisis information: An analysis of official sources during the 2014 Carlton complex wildfire. In *Proceedings of the 35th International Conference on Human Factors in Computing Systems* (CHI 2017). Retrieved from: https://dx.doi.org/10.1145/3025453.3025627

De Fina, A., & Georgakopoulou, A. (2012). *Analyzing narrative: Discourse and sociolinguistic perspectives*. New York: Cambridge University Press.

Derks, D., & Bakker, A. B. (2014). Smartphone use, work–home interference, and burnout: A diary study on the role of recovery. *Applied Psychology, 63*, 411–440. https://dx.doi.org/10.1111/j.1464-0597.2012.00530.x

East, S. (2016, August 1). Teens: This is how social media affects your brain. *CNN.com*. Retrieved from: www.cnn.com/2016/07/12/health/social-media-brain/index.html

Endsley, T., Wu, Y., Reep, J., Eep, J., & Reep, J. (2014). The source of the story: Evaluating the credibility of crisis information sources. In S. R. Hiltz, M. S. Pfaff, L. Plotnick, & P. C. Shih (Eds.), *Proceedings of the Information Systems for Crisis Response and Management* (ISCRAM) (Vol. 1, pp. 158–162). University Park, PA: ISCRAM.

Faraj, S., Jarvenpaa, S. L., & Majchrzak, A. (2011). Knowledge collaboration in online communities. *Organization Science, 22*, 1224–1239. https://dx.doi.org/10.1287/orsc.1100.0614

Gao, W., Liu, Z., Guo, Q., & Li, X. (2018). The dark side of ubiquitous connectivity in smartphone-based SNS: An integrated model from information perspective. *Computers in Human Behavior, 84*, 185–193. https://dx.doi.org/10.1016/j.chb.2018.02.023

Georgakopoulou, A. (2017). Sharing the moment as small stories: The interplay between practices & affordances in the social media-curation of lives. *Narrative Inquiry, 27*, 311–333. https://doi.org/10.1075/ni.27.2.06geo

Goffman, E. (1959). *The presentation of self in everyday life.* New York, NY: Doubleday.

Hutchby, I. (2014). Communicative affordances and participation frameworks in mediated interaction. *Journal of Pragmatics, 72*, 86–89. https://dx.doi.org/10.1016/j.pragma.2014.08.012

Hughes, D., McGillivray, L., & Schmidek, M. (1997). *Guide to narrative language: Procedures for assessment.* Eau Claire, WI: Thinking.

Jin, Y., Liu, B. F., & Austin, L. L. (2011). Examining the role of social media in effective crisis management: The effects of crisis origin, information form, and source on publics' crisis responses. *Communication Research, 41*, 74–94. https://dx.doi.org/10.1177/0093650211423918

Kreps, G. L. (2009). Applying Weick's Model of Organizing to health care and health promotion: Highlighting the central role of health communication. *Patient Education and Counseling, 74*, 347–355. https://doi.org/10.1016/j.pec.2008.12.002

Labov, W. (1972). *Language in the inner city.* Philadelphia: University of Pennsylvania Press.

Leonardi, P. M., & Vaast, E. (2017). Social media and their affordances for organizing: A review and agenda for research. *Academy of Management Annals, 11*, 150–188. https://doi.org/10.5465/annals.2015.0144

Lepp, A., Li, J., Barkley, J. E., & Salehi-Esfahani, S. (2015). Exploring the relationships between college students' cell phone use, personality, and leisure. *Computers in Human Behavior, 43*, 210–219. https://dx.doi.org/10.1016/j.chb.2014.11.006

Maitlis, S., & Sonenshein, S. (2010). Sensemaking in crisis and change: Inspiration and insights from Weick (1988). *Journal of Management Studies, 47*, 551–580. https://dx.doi.org/10.1111/j.1467-6486.2010.00908.x

Meyers, K. K., & Sadaghiani, K. (2010). Millennials in the workplace: A communication perspective on millennials' organizational relationships and performance. *Journal of Business and Psychology, 25*, 225–238. https://dx.doi.org/10.1007/s10869-010-9172-7

Oulasvirta, A., Rattenbury, T., Ma, L., & Raita, E. (2011). Habits make smartphone use more pervasive. *Personal and Ubiquitous Computing, 16*, 105–114. https://dx.doi.org/10.1007/s00779-011-0412-2

Palen, L., Vieweg, S., Liu, S. B., & Hughes, A. L. (2009). Crisis in a networked world: Features of computer-mediated communication in the April 16, 2007, Virginia Tech event. *Social Science Computer Review, 27*, 467–480. https://dx.doi.org/10.1177/0894439309332302

Pipino, L., Lee, Y. W., & Wang, R. Y. (2002). Data quality assessment. *Communications of the ACM, 45*(4), 211–218. https://dl.acm.org/citation.cfm?doid=505248.506010

Reynolds, N., & Seeger, M. W. (2005). Crisis and emergency risk communication as an integrative model. *Journal of Health Communication, 10*, 43–55. https://dx.doi.org/10.1080/10810730590904571

Rosile, G. A., Boje, D. M., Carlon, D. M., Downs, A., & Saylors, R. (2013). Storytelling diamond: An antenarrative integration of the six facets of storytelling in organization research design. *Organizational Research Methods, 16*, 557–580. https:// dx.doi.org/10.1177/1094428113482490

Samaha, M., & Hawi, N. S. (2016). Relationships among smartphone addiction, stress, academic performance, and satisfaction with life. *Computers in Human Behavior, 57*, 321–325. https://dx.doi.org/10.1016/j.chb.2015.12.045

Sobkowicz, O., Kaschesky, M., & Bouchard, G. (2012). Opinion mining in social media: Modeling, simulating, and forecasting political opinions in the web. *Government Information Quarterly, 29*, 470–479. https://dx.doi.org/10.1016/j.giq.2012.06.005

Sorensen, J. H. (2000). Hazard warning systems: Review of 20 years of progress. *Natural Hazards Review, 1*(2), 119–125. https://ascelibrary.org/doi/10.1061/%28ASCE% 291527-6988%282000%291%3A2%28119%29

Stephens, K. K. (2018). *Negotiating control: Organizations and mobile communication.* New York, NY: Oxford University Press.

Stephens, K. K., Barrett, A. K., & Mahometa, M. L. (2013). Organizational communication in emergencies: Using multiple channels and sources to combat noise and capture attention. *Human Communication Research, 39*, 230–251. https://dx.doi. org/10.1111/hcre.12002

Stuart, H. C., Dabbish, L., Kiesler, S., Kinnaird, P., & Kang, R. (2012). Social transparency in networked information exchange: a theoretical framework. *Proceedings of the ACM 2012 Conference on Computer Supported Cooperative Work, Seattle, Washington,* 451–460. https://dx.doi.org/10.1145/2145204.2145275

Sutton, J., Palen, L., & Shklovski, I. (2008). Backchannels on the front lines: Emergent uses of social media in the 2007 southern California wildfires. *Proceedings of the 5th International ISCRAM Conference*, Washington, DC. Retrieved from: www.iscram.org/legacy/dmdocuments/ISCRAM2008/papers/ISCRAM2008_ Sutton_etal.pdf

Veil, S. R., Buehner, T., & Palenchar, M. J. (2011). A work-in-process literature review: Incorporating social media in risk and crisis communication. *Journal of Contingencies and Crisis Management, 19*, 110–122. https://dx.doi. org/10.1111/j.1468-5973.2011.00639.x

Weick, K. E. (1979). *The social psychology of organizing.* Reading, MA: Addison Wesley.

Weick, K. E. (1988). Enacted sensemaking in crisis situations. *Journal of Management Studies, 25*, 305–317. https://dx.doi.org/10.1111/j.1467-6486.1988.tb00039.x

Weick, K. E. (2007). The generative properties of richness. *Academy of Management Journal, 50*, 14–19. https://dx.doi.org/10.5465/amj.2007.24160637

Weick, K. E., Sutcliffe, K. M., & Obstfeld, D. (2005). Organizing and the process of sensemaking. *Organization Science, 16*, 409–421. https://dx.doi.org/10.1287/ orsc.1050.0133

White, C.M. (2012), *Social media, crisis communications and emergency management: Leveraging Web 2.0 technology.* New York, NY: Taylor & Francis Group.

Wilensky, H. (2014). Twitter as a navigator for stranded commuters during the Great East Japan earthquake. In S. R. Hiltz, L. Plotnick, M. Pfaf, & P. C. Shih (Eds.), *Proceedings of the Information Systems for Crisis Response and Management (ISCRAM)* (pp. 695–704). University Park, PA: ISCRAM.

2

THIS IS GETTING BAD

Embodied Sensemaking about Hazards When Business-as-Usual Turns into an Emergency

Jody L. S. Jahn

He's like *should we go through [the fire]?* And I was like, *we don't have a choice.* Let's get moving or we're going to be crispy critters out here…[A]s soon as I moved, flames just went WHOOSH right by me. I held my breath as I ran through the flaming front, because you'll burn your lungs pretty easy, and then just kept running as far as I could. And I just remember thinking *this is really hot, and, I don't want to, but I need to breathe…* The fire went so fast—rabbits and coyotes and stuff that were partially burned were kind of running around out there. And we were out there, too, with the rest of the burned animals, which was really weird, you know, just to see.

Ben, 15 wildland firefighting seasons

Favorite quote of mine is "experience is something you gain right after you need it," and sometimes, some things in fire you need to experience to learn how to deal with it.

Trevor, nine wildland firefighting seasons

Managing emergencies and hazards is business-as-usual for members of high-reliability organizations (HROs) in which members commonly face thin margins of error, fast work tempos, and dangerous conditions (Weick & Sutcliffe, 2015). Because accepting some amount of risk is part of the job, HRO members can easily become desensitized to the hazards they regularly face, and they might succumb to pressure to accept risks beyond what they feel comfortable taking on. Because members' lives are at risk, it is important to know when a "normal" hazard becomes a threat. This study explores wildland firefighter sensemaking about "close-calls" in which they narrowly escaped life-threatening firefighting situations that began as business-as-usual, and the social sensemaking processes that helped them learn from those experiences. The goal of this study is to probe how members make sense of the liminal period of time between when a hazard becomes a threat, and how their sensemaking about these experiences contributes to their expertise in managing the complex, emerging, ambiguous circumstances that HROs commonly face.

HROs are a kind of organization that operates in high-tempo situations with minimal margins of error (Rochlin, 1993; Weick & Sutcliffe, 2015). There is typically some source of complexity or ambiguity that can pose dangers, and which members must closely monitor. For example, air traffic controllers arrange the order and timing of numerous aircraft while avoiding collisions (Weick & Sutcliffe, 2015). Emergency medical teams sort through a patchwork of symptoms to make a diagnosis (Blatt, Christianson, Sutcliffe, & Rosenthal, 2006). Nuclear power plants have numerous lights and dials, and complex mechanics making it difficult to identify small problems before they get big. Wildland firefighters, the focus of this study, face ambiguity from the wildfire environment (Barton & Sutcliffe, 2009; Jahn, 2016). If HRO members commonly face tricky and uncertain conditions, how can they know when a *dangerous* situation becomes a *threatening* one? One explanation is that they learn from close-call situations, accidents, and fatalities to make distinctions between hazardous and threatening situations.

Hazards, Threats, and Risk Assessment

What is the difference between a hazard and a threat? A *hazard* exposes one to chance—to the possibility of injury if the hazard is not addressed ("hazard," n.d.), while a *threat* places a person in a compelling situation in which their life is at stake ("threat," n.d.). Not all hazards are threatening, but threatening situations always contain hazards. Whether a hazard turns into a threat depends on how vulnerable somebody is to the hazard. A great deal of risk assessment literature looks at how organizations and individuals identify and plan around hazards in order to avoid threatening situations (see van Duijne, van Aken, & Schouten, 2008). There are rules, guidelines, and mathematic probability models that can predict the likelihood of threats given certain hazards. Risk assessment literature also is concerned with the complexities of decision-making in the midst of hazards (see Aven, 2009).

While these rules, mathematic probability models, and decision-making premises play a role in assessing hazardous environments, it also is important to develop our understanding of how workers learn to make distinctions between hazards and threats through their lived, embodied experiences of encountering them, making mistakes, getting injured, and through narrowly escaping injuries and death. Lived experiences become important occasions for sensemaking, producing narratives through which members construct cause, effect, and important lessons that inform how they plan to deal with similar situations in the future.

Sensemaking in Response to Cosmology Events

The notion of *sensemaking* can help us understand how professionals and organizations make sense of differences between hazards and threats (Weick, 1995). Sensemaking refers to a process of action and reflection, in which organizational actors retrospectively devise an explanation for the meaning of something they have previously enacted or experienced (Maitlis & Christianson, 2014; Weick, 1995). Often sensemaking is prompted by surprises or unexpected turns of events. Karl Weick (1993) used the term "cosmology event" to describe situations that are no longer sensible; that is, situations in which the complexity or order of events ceases to be explainable or comprehendible. Weick analyzed Norman Maclean's (1992) famous account of the 1949 Mann Gulch wildfire, which killed 13 firefighters. A centerpiece of Weick's (1993) sensemaking analysis was a survivor's account of a cosmology event in which the fire grew large so quickly that it defied his ability to anticipate what to do next. Then—in the heat of the moment—he improvised an ingenious solution that saved his life.

Cosmology events are key moments of surprise that prompt a search for meaning and an explanation of what occurred (Weick, 1993). Weick (1995) came to refer to such events as *occasions for sensemaking*, while other authors have variously referred to them as shocks, surprises, and crises that prompt members to ask "what is going on here?" When faced with a cosmology event, members bracket important cues from the situation, and craft a narrative that makes both actions and events sensible (Weick, 1995). The process of defining the situation generates a narrative, account, or explanation that the person then carries forward to future situations. A cosmology event can be considered an extreme occasion for sensemaking. Cosmology events are important because they can help us to better understand embodied sensemaking, in particular: (a) how learning processes are rooted in spatial materiality of hazardous environments, and the physicality of the body within those spaces, and (b) how sensemaking involves interpretations of both environmental cues and other members' responses.

Embodied Sensemaking

Cosmology events provide insights into ways that sensemaking processes are corporeal, material, and spatial (Hultin & Mähring, 2017; Mills, 2002; Whiteman & Cooper, 2011). For example, examining member sensemaking about close calls can help us understand how organizational knowledge (e.g., codified lessons written in fatality and accident reports) is translated into action in the ambiguous, emerging environments in which HROs regularly operate. One example of organizational knowledge relevant to HROs includes these organizations' safety rules. In many HROs, safety rules are lessons extracted from accidents and fatalities that have been codified into a written form (Jahn, 2016; Sauer, 1998; Ziegler, 2007). However, a problem with safety rules is that they are either written in abstract language, so that they can apply to any situation, or they are so specific that they do not assist members in discerning nuances in their present circumstances (Hale & Borys, 2013; Sauer, 1998). Yet, the lessons embedded in the rules still carry life-and-death importance. Rules require a translation from a written modality to a lived modality (Sauer, 1998), a translation that likely occurs through *enactment* (Weick, 1979).

Sensemaking involves enactment of spaces, meaning that members take action in their physical environments through moving their bodies across landscapes, implementing tasks in terrain, arranging and rearranging material objects in space, and other activities. For instance, in Weick's (1993) analysis of the Mann Gulch incident, a firefighter threatened with a rapidly approaching flame front made the counterintuitive decision to set on fire a patch of grass in his immediate area. This was an example of enacting a physical environment; the speed of the approaching flame front did not make sense to the firefighter and was the source of his cosmology event. However, what *did* make sense in that moment was the fact that, once the grass around him has burned, it would not burn again. His working theory in the moment was that a consumed patch of grass would protect him by slowing the speed of the fire in that small area. The firefighter enacted his environment by burning the grass before the flame front arrived, thus testing his working theory, and arriving at an account of how to handle his physical environment in future similar circumstances. The firefighter's actions were an example of enactment because he bracketed environmental cues and made sense through moving his corporeal body through the landscape and taking action on the physical terrain around him.

Examining sensemaking about cosmology events can also help us understand how sensemaking involves the enactment of social systems, meaning that it takes place amid social groups, organizational expectations, and cultural practices (Roberts, Rousseau, & La Porte, 1994). Sensemaking is social in a variety of ways, namely, through frames of reference that shape what and how people see (Brummans et al., 2008); collective sensemaking with others (Wolbers & Boersma, 2013); and individuals' narratives that explain both an event and how they saw others responding to it (Cunliffe & Coupland, 2012). Organizations are social systems in which members' actions are interdependent. In particular, members experience their interdependence through "double interacts" involving action, reaction, and feedback to the reaction (Weick, 2004). In other words, one develops an initial understanding of a situation, reads it against others' interpretations or visible reactions, then revises or solidifies their own understanding. Cues from fellow members (e.g., their nonverbal messages, body movements, facial expressions, tone of voice, etc.) importantly mediate whether and how people interpret the potential and severity of emerging situations. Weick's theorizing about "double interacts" can be applied to thinking about how members' bodies cue others' interpretations of circumstances to contribute to intercorporeal knowing and improvisation.

Cosmology events involve certain kinds of performances of organizational codified knowledge, spaces, and intercorporeal knowing. *Intercorporeal knowing* is defined as a sensitivity to the "delicate and subtle shifts in the embodied conduct of colleagues" (Hindmarsh & Pilnick, 2007, p. 1413). Hindmarsh and Pilnick contend that intercorporeal knowing involves understanding how current movements indicate trajectories of action, and opportunities to assist in task accomplishment. This kind of intercorporeal communication is similar to the improvisational interactions characteristic of jazz musicians (Miner, Bassoff & Moorman, 2001; Weick, 2006). In jazz, improvisation involves extemporaneously composing an original piece of music through coordinating the activities of ensemble members whose music becomes an elaboration of pre-composed musical phrases (Miner et al., 2001). Improvisation does not occur spontaneously; rather it is rooted in common understandings of simple, minimal structures (Mendonça, e Cunha, Kaivo-oja, & Ruff, 2004; Weick, 2006). Applied to organizations, it is important to note that improvisation depends on sets of rules that guide action, fragments of routines, and trajectories of action. Improvisational performances require that all members have some knowledge of the basic rules, that the more experienced members are sensitized to the experience levels of less practiced members, and that they socialize members toward a deeper awareness of the group's conventions. Improvisational performances can afford a greater repertoire of flexible actions anchored in the crew's functional routines and notions of appropriate actions. Improvisational performances arise as HRO members anticipate potential failures and readjust their actions in real time. The guiding research question for this study is as follows:

RQ: How do organization members narratively construct close-call experiences into *sense*-able lessons?

Method

Wildland firefighting is considered an exemplary high-reliability organization (Barton & Sutcliffe, 2009; Jahn, 2016; Weick & Sutcliffe, 2015), with the U.S. Departments of Agriculture and Interior employing approximately 7,000 full-time wildland firefighters, and thousands more seasonal firefighters.

I spent eight seasons as a wildland firefighter. The reasons why I chose that line of work related as much to my upbringing as they did to my interest in the work. My father spent a considerable portion of his Forest Service career as a field-going forest ranger. As a result, I spent my childhood in rural communities in which the Forest Service was among the few large employers. Growing up, I witnessed my father's personal commitment to his work and to the organization, much like Kaufman (1960) described in his study on the forest ranger. At my father's suggestion, I began working for the Forest Service the day after my eighteenth birthday. My first job was on a wilderness trail crew that was also trained for wildland firefighting. That season was my entrée into the profession and thereafter I pursued work on crews whose sole purpose was firefighting. By the time I left fire to pursue graduate school, I had spent at least some span of time working in every wildland fire specialty except for smokejumping. I worked on an engine, a type 2 handcrew, a hotshot crew, and my fire career culminated with three seasons in a permanent federal appointment as a mid-level supervisor on a helicopter rappel crew. Gaining experience throughout the organization enabled me to experience firsthand both the dynamic firefighting environment and the complexity of the organizing processes that occur in response to it. These experiences stimulated my interest in studying the profession.

Data Collection

This chapter analyzes interviews from 37 firefighters from two crews that specialize in helicopter rappel operations. Overall, interviews lasted between 30 to 60 minutes, and were recorded, transcribed, and labeled with each participant's pseudonym. Interviews yielded more than 400 pages of single-spaced transcript. Interviews were conducted on site at each crew's base station. To ensure privacy, each interview took place in a private office with the door closed. After gathering information from participants related to demographics and his or her firefighting background, I followed a semi-structured format. To ascertain participants' cosmology events, I asked them to tell a critical incident story (Flanagan, 1954; Gremler, 2004). A *critical incident* is an account of activities observed by the participant and complete enough to enable the researcher to make inferences about the actors involved in the actions (Gremler, 2004). I asked each participant to talk about a memorable fire experience that he or she felt was important for developing his or her expertise as a firefighter. Several participants immediately came up with stories to tell while other members struggled to recall one. To cue their memories, I further explained that I was interested in experiences such as times when they were surprised by fire activity, took on a position of responsibility or leadership for the first time, experiences when something when wrong, or times when an incident went particularly well.

Data Analysis

I first read through all of the interviews to get a feel for participants' responses and to get an overall idea of the content of their critical incident stories. During this initial review, I extracted all of the critical incident accounts. Second, I systematically coded each transcript in its entirety, including the critical incidents. To develop the initial coding dictionary, I selected four transcripts from participants who had worked on their respective crew for more than one season, and who stood out as descriptive and insightful about their experiences. The majority of codes for the dictionary were captured from these rather rich interviews. Once the coding dictionary was developed, I coded all the interviews. I updated the coding dictionary after each subsequent transcript to reflect new codes that emerged and to refine coding categories. I used an open coding procedure (Tracy, 2012), which involved reading through the critical incidents line by line—sometimes paragraph by paragraph—to label what participants discussed about their embodied sensemaking from their critical incidents.

Finally, I sorted the critical incidents based on similarities regarding the types of events that occurred in the story, and the ways members talked about making sense of what had occurred. There were three categories of stories, which I later labeled as *instantiating performances*, *failure performances*, and *improvisational performances*. Within the three categories of stories, I examined how bodily enactments of close-call (and other) situations seemed to play a role in

TABLE 2.1 Embodied sensemaking performances

	Instantiating performances	*Failure performances*	*Improvisational performances*
How sense is made	Enactment under uncertain or threatening circumstances prompts a translation of knowledge from a written to a lived modality.	A failure or close call prompts a reorganization or reinterpretation of codified knowledge or physical environments member thought s/he already knew.	Members read physical environments and trajectories of action through a social lens. Double interacts help us understand how improvisation is rooted in social processes.
Narrative structure of members' sensemaking	Initial understanding of codified knowledge (e.g., safety rule). Enacting codified knowledge in uncertain/threatening situation. New understanding of codified knowledge resulting from appropriating and physically enacting it.	Acting knowingly/ skillfully in a familiar environment or using familiar practices. Experiencing failure *or* utilizing a last resort contingency. Reorganizing one's understanding of an environment or practice. This narrative structure also applies to members witnessing somebody else's failure.	Members interpret what they see in their environment. Members then look to see how others are responding to the situation through recognizable repertoires of verbal, nonverbal, kinesthetic, and/or vocalic phrases indicating "normal" or "abnormal" circumstances. They then solidify or alter their interpretation.
What sense is made from cosmology event	More nuanced, embodied understanding of organizational lessons codified in safety rules.	Clearer understanding of one's limits. Less equivocal, more resolute voice.	Greater sensitivity to reading meaningful "minimal structures" of verbal, nonverbal, and kinesthetic behavior.

how members came to better understand both routine and unusual fire situations, how those enactments were connected to organizational knowledge (e.g., safety rules, checklists, etc.), and how members interpreted their physical performances in relation to ways they witnessed fellow employees responding to shared situations. The three categories of stories indicated distinctive physical performances, each with a different narrative structure, and they form the basis for the findings (see also Table 2.1).

Findings

The central argument in this study is that cosmology events prompt a particular kind of sensemaking that occurs through bodily enactment (Weick, 1979), and is codified into a narrative by which the individual makes sense of the event. The findings identify three sensemaking performances by which organization members respond to cosmology events: instantiating performances, failure performances, and improvisational performances. Each of these embodied sensemaking performances was simultaneously material and social—embedded both in the material environments in which members were facing bodily threat, and in social interactions in which they assessed the level of threat, in part by paying attention to how people around them were responding to the situation.

Instantiating Performances

Instantiating performances occurred when members acted out what they had learned outside of an actual situation in a way that cemented the insights into embodied knowledge. This might have occurred as a member acted out in real life what he or she had learned in training, or by encountering a surprise or counterintuitive experience. Instantiating performances were instances of enactment under uncertain or threatening circumstances that prompted a translation of knowledge from a written to a lived modality. For example, a wildland firefighter might have learned from classroom training that large accidents often are the result of small errors that, if ignored, accumulate and "line up" like holes in Swiss cheese (Reason, 1998). Each defense against a potential failure is a slice of Swiss cheese, and the weaknesses in the defense are represented by the holes in the cheese. As weaknesses such as gaps in supervision and lack of experience accumulate, the holes in the Swiss cheese align, and hazards breech the system's defenses (Weick, 2003). To prevent failures from happening, fire fighters are taught to enact practices that provide the best possible defense (e.g., appointing lookouts to monitor fire behavior and changing weather conditions). However, it is through fireline enactments that they come to understand how some defenses are associated with avoiding particular types of failures, and when some defenses are more appropriate than others.

Jean described a close call in which her crew was surprised by fire. When discussing the fire later in an after action review (AAR), a learning practice, Jean noted how the close call made salient several common safety rules and guidelines. Indeed, the intensity of the situation emphasized the importance of several safety guidelines, and deepened her understanding of what the safety measures intended to accomplish:

> Jean: We had an AAR with just the crew and I remember telling them "look the fact that we got out of there safely and we're OK is awesome because now we have this learning experience—you guys are all ten times better firefighters right now than you were before you went to that fire… You guys have all the things you learned in class, all the Watchouts and Fire Orders—so many of those came into play on that fire." For me it was a big wake up to realize, yeah, I've reviewed all that stuff in class: [various checklists], this, that, you know, Swiss cheese lining up. Well I got to see it *actually happen* there and you know it was like, OK, this stuff isn't just something we go over in class to check off that I have this class. This helps you out in the field for sure, and yeah, it absolutely affects the decisions I make now.

As demonstrated in Jean's account, instantiating performances followed a narrative structure (Table 2.1) that involved an initial understanding of codified knowledge (e.g., a safety rule). The participant then described enacting codified knowledge in an uncertain or threatening situation. From the enactment, the organization member came to a new understanding of codified knowledge resulting from appropriating and physically enacting it. The result of the sensemaking process was a more nuanced and embodied understanding of the organizational lessons codified in safety rules and checklists.

Not all instantiating performances occurred under duress. Several interviewees described stressful events in which they were in charge for the first time, particularly referencing the rules and guidelines as they made decisions. Whether facing duress or everyday performance pressure, the narrative structure was the same. For example, Stan described an instantiating performance in which he was in charge of coming up with a tactical firefighting plan for the first time at the request of a higher-ranking firefighter, whom he respected and wanted to impress:

> Stan: This guy hiked my tail off all day looking, checking there, making sure everything was good. And then… he just says: *so what should we do? What do you think?*… I was like, dang it, I knew he was going to ask me that, you know… And so I just thought about it… and I just said, obvious kind of conclusion is to bring our folks down from the bottom and start up from the wet drainage. Downhill line construction checklist [came to

mind], that whole stuff. Started the wires arcing, bzz, made a little connection there… it was kind of cool for me, because it was, you know, it put the pressure on me to kind of come up with a solution and I thought it went pretty good. And so it was kind of cool to have somebody else from another crew… let me call the shots and actually empowered me to make the decisions.

For instantiating performances, the differentiation between hazard and threat came from interpreting textbook or rule-based knowledge in light of lived experiences, what Sauer (2003) referred to as translating knowledge from a written modality to an embodied or lived modality. A second embodied sensemaking performance was a failure performance.

Failure Performances

While an instantiating performance involved transferring knowledge from a written to a lived modality, a failure performance was a reorganization or reinterpretation of knowledge or physical environments the member already thought they knew. Here, the enactment process resulted in a deeper knowledge that could be defended with a stronger, more resolute voice. Weick and Sutcliffe (2015) have discussed the importance of failures in HROs. Preoccupation with failure means that organizations are sensitive to as many lessons as possible from failure and near-failure incidents. The failures that seem far removed from personal experience may not affect everyone in meaningful ways when they are not experienced firsthand. Failure performances arise not just from being surprised, but specifically by having to *enact* the worst-case scenario. For example, wildland firefighters enact failure performances when utilizing a safety zone, deploying the fire shelter, dropping tools or gear to enable escape, or running away from fire when it is normal to walk. Through failure performances, the body goes through the characteristic sensory experiences involved in a life-threatening failure, regardless of the actual outcome (i.e., whether there was actually a failure). The body becomes more knowledgeable as it performs the motions of failure, specifically experiencing the shock and emotions of what it is like leading up to a total failure.

For example, wildland firefighters are taught to always have a safety zone nearby in the event that they face a worst-case scenario. A firefighter may have an understanding about the approximate size of a safety zone and the permissible amount of vegetation that still allows for safety based on training and enacting routine practices. The firefighter may also understand the appropriate distance of and type of terrain present in an escape route that connects working firefighters to the safety zone. However, this rational, conceptual knowledge may become truly relevant or *sensible* only after the firefighter has had to actually hike to the safety zone under stress. Thus, enacting the escape route and safety zone is a failure performance in which the body labors across terrain, experiencing the physical reality of the escape route, and thereby coming to understand this already-held knowledge about escape routes and safety zones in a new way (e.g., escape routes would be safer if shorter, closer, not as steep, etc.). From the enactment of failure, the body becomes "wise," recognizing the limits and material realities of what is considered taken-for-granted and normative firefighter knowledge.

Sean described a fire event in which he served in a command position when firefighters had died from being overrun by fire, or *burned over*. He described how the event changed what he considered to be an "acceptable risk." Sean said incidents like the one he experienced made people into better firefighters because, through being "bitten," "burned," or "shown," firefighters became acutely aware of their limits in a way inaccessible to them before such an incident.

> Sean: I guess part of my story is that my definition of what a safety zone is has definitely changed because of this experience... And it made [me] look at things from several different angles... I think if people haven't had those experiences, you know, those close calls or those "oh shit" moments you know... if people haven't been burned—not literally—but if you haven't had a close call or seen some close calls, then it changes what you feel is an acceptable risk... Things that now make me leery are... people who I know haven't been bitten or shown. That concerns me because I do see them acting more aggressively toward fire... You just realize that they haven't had that [failure] experience quite yet and they haven't had their decisions change because of it... I'm always a little leery if they don't know their limits.

A failure or close call prompted a reorganization or reinterpretation of codified knowledge or physical environments member thought s/he already knew. Failure performances followed a narrative structure that included acting knowingly or skillfully in a familiar environment or using familiar practices, experiencing failure (or utilizing a last-resort contingency), then engaging in sensemaking that reorganized his or her understanding of an environment or practice. In addition to developing a clearer understanding of one's limits, an important result of failure performances was that it helped members develop a more resolute voice, particularly a dissenting voice. Both Travis and Amy described how failure performances contributed to their development of a stronger dissenting voice:

> Travis: One time our crew nearly got burned over and a few people got— they got small burns and stuff... Everyone was frustrated with the trainee incident commander of the fire at the time... And they said they could all

sense it and feel it coming. He was having them hold the fireline mid-hill and [burned material] was rolling [below them on the hill], then fire just blew over the top of them. I guess a lot of times you can just sense it, just by the commands you're getting. It just doesn't jive with how you would do it, but then again, they're in charge. It's hard to make a fuss because a lot of times it will work out; it's not always a perfect situation... [But now] if I feel something funny about the leadership, I'm not going to give them any slack, you know.

Amy: And I think I even said it to my squad boss [when discussing the incident later] I think all it would have taken would have been one person to say *no, I'm not going down the hill.* If there had just been one person to sort of throw a wrench into things, that would have shut everything down, because then they would have to stop. They would have had to talk to the person. Then they would have called down and told our superintendent [that] we're getting a refusal up here... So just more of turning it into action: not just saying I don't *think* we should go, but I'm *not going* to go.

Failure performances differed from instantiating performances in that instantiating performances enabled members to translate knowledge from a textual modality to a lived or physical one, while failure performances involved a reconfiguration or reinterpretation of what an organization member thought she or he knew, or resulted in a change in their usual method for approaching or handling a situation. Failure performances contributed to helping firefighters make distinctions between hazards and threats through making limits clear, clarifying the need for voice, and emboldening firefighters to use their voice (i.e., to "speak up) in the future.

Improvisational Performances

Improvisation is ultimately about working together toward creative action through utilizing minimal structures. Part of improvisation is members understanding the trajectory of action *so far* so they can figure out what to do next. Improvisational performances enabled members to discern the difference between hazard and threat through ways that in-the-moment creative action under duress prompted them to notice key aspects of situations that were entirely unknown, unseen, or deemed unimportant before the incident. Following theorizing about improvisational jazz, improvisational performances were inherently social, as members read physical environments and trajectories of action through a social lens. In particular, members' interpretation aligned with Weick's (2004) notion of a "double interact." Double interacts involve action, reaction, and feedback to the reaction (Weick, 2004). In other words, one develops an initial understanding of a situation, reads it against others' interpretations, then revises or solidifies his or her own understanding. Cues from fellow members (e.g., nonverbal messages, body movements, etc.) importantly mediate whether and how people interpret the potential and severity of emerging situations. For instance, Paul described a fire situation in which the fire conditions changed. While Paul did not see the change in fire conditions as requiring a sudden change of action, he witness a co-worker, who had a similar amount of experience as he did, react in a frantic manner that forced him to question his observations about the situation. He reflected on the experience, saying:

> Paul: When somebody else on your crew has the same experience or has more experience than you and they are visibly upset or completely going bonkers, like "we gotta get out of here!" and they're running through the bushes and that kind of thing, it makes you wonder "what am I seeing?" It's hard to actually pay attention to your own feelings at that time because they're clouded by, I think, other people's visions, too. You're thinking "hey, maybe I'm not seeing something right, maybe somebody else has experiences that I don't have."

Paul's account pointed to a non-normative way in which his crew member physically enacted the fire environment, running through the bushes, "going bonkers," and looking visibly upset. The co-worker's actions were unusual in an environment in which firefighters move their bodies deliberately (rather than hurriedly), and remain calmly watchful (rather than upset). Not only did the co-worker's actions influence Paul's evaluation of the fire environment, also important was the weight of the co-worker's actions given that Paul saw that he and his co-worker shared similar levels of fire experience. The co-worker's actions carried such a weight that Paul paused to reevaluate what he was seeing in the fire environment. Ultimately, Paul determined that the situation was serious enough to move to a safer place, but he emphasized that his peer's response to the situation was unnecessarily frantic.

During a cosmology event, the environment or situation becomes unrecognizable. Because the usual expectations did not apply, members looked elsewhere to decide how to interpret what they were seeing and experiencing. A variety of physicality factors may go into an overall assessment of another person within confusing circumstances. For instance, the complementary relationship between a supervisor and a subordinate might afford weighted interpretations of one another's actions such that when the boss' behaviors (e.g., rushed movements, sweating, shaking) show uneasiness about a situation, the display might carry more gravity than if a less experienced subordinate displayed the same behaviors. Thus, the interaction of organization member's salient identities in the moment, combined with the physical actions and behaviors displayed, might shape the attributions made about their behaviors and cues, and interpretations of them. To illustrate, Robin described an incident in which she was marking an escape route (a common practice) that she and others ended up using right away to escape the fire (a relatively uncommon occurrence):

> Robin: When it first hit home like "this is big" was when I was marking out the escape route because I was the slowest person on the crew [laughs] and I was like "wow they're really gonna use this." The most intense part was really when [the supervisor] grabbed my shoulder and we left the fireline… There was no room for any type of conversation. He said, "OK, let's go." It was very short and clipped. I couldn't sense any worry. It was just matter of factly—"OK, we're going." And I just knew it was serious. It was time to go.

In improvisational performances, members interpreted what they saw in their environment, they then looked to see how others were responding to the situation through recognizable repertoires of verbal, nonverbal, kinesthetic, and/ or vocalic phrases indicating "normal" or "abnormal" circumstances, and they finally solidified or altered their interpretations. The sensemaking result of improvisational performances was a greater sensitivity to reading meaningful "minimal structures" of verbal, nonverbal, and kinesthetic behavior.

Discussion

The main goal of this study was to probe how HRO members make sense of the liminal period of time between when a hazard becomes a threat, and how their sensemaking about these experiences contributes to their expertise in managing complex, emerging, ambiguous circumstances. The study contributes to our understanding of embodied sensemaking in three important ways. First, it helps us better see how sensemaking is a social and corporeal process. Second, it exposes how embodied sensemaking is involved in the development of professional expertise. Third, it suggests ways that members in an expertise-driven occupation develop a voice for proposing alternatives and offering dissenting views.

Sensemaking as a Social and Corporeal Process

This study extends what we know about enactment by grounding it in physical performances of instantiating lessons and knowledge, in performing failure, and in spontaneous acts of improvisation—particularly double interacts. Weick often talks about enactment and action as fodder for interpretation and meaning (Weick, 1995). This study concurs with such a take on action, but differentiates between types of sense that arise from various physical performances of organizational knowledge and physical space. Specifically, this study extends Whiteman and Cooper's (2011) work on ecological sensemaking, which focuses on physical enactments of spaces and circumstances in the development of deep knowledge about the nuances of one's environment. However, this study includes a broader view of sensemaking that involves not just physical enactment in space, but also incorporates the spatial performance of organizational knowledge (i.e., safety rules) into the sensemaking process. This is an important development because organizational action is inherently informed by organizational texts.

The Role of Sensemaking in Developing Expertise

By centering on spatially enacted performances, this study's findings help us understand how sensemaking is involved in the development of expertise. In particular, these findings illustrate how sensemaking and expertise arise from different types of physical performances: instantiation, failure, and improvisation. Considering embodied performances directs attention to ways that lived experiences provide both narratives and meaningful enactments that members use as a basis for comparing their experience to that of others in expertise-driven organizations. Having a particular kind of narrative to tell or invoke provides members with both cultural capital and evidence to justify decisions and actions. These findings complement Ishak and Williams' (2017) study examining how wildland firefighters "borrow" each other's visceral experiences through taking on aspects of those narratives as their own. Thus, considering the present study's findings in relation to Ishak and Williams' study highlights the importance of organizational practices that allow for the sharing of professionals' experiences so that they can learn important, subtle, visceral, and crucial lessons from each other.

Gaining a Voice in an Expertise-Driven Occupation

A few sensemaking studies in HRO contexts have examined how expertise enables and constrains professionals' voice (Barton & Sutcliffe, 2009; Blatt et al., 2006; Krieger, 2005; Lewis, Hall, & Black, 2011). This study adds to our understanding of voice in HROs by anchoring voice in lived experiences, which adds nuance to studies that emphasize the discretionary nature of voice without considering how embodied performances inform the resoluteness with which members voice concerns and dissent. High-reliability organizations require members to communicate to construct common understandings of an environment or emerging event, and to coordinate activities (Rochlin, 1993; Weick & Sutcliffe, 2015). Voice is crucial in HRO contexts because constantly managing dangerous and fluctuating situations requires members to exchange information, voice assent and dissent, and propose alternative plans and viewpoints without hindrance (Krieger, 2005; Lewis et al., 2011). These voicing activities disrupt what Barton and Sutcliffe (2009) refer to as *dysfunctional momentum*, or problematic trajectories of action. When members voice concerns, their statements remain in the air, prompting fellow members to choose whether and how to alter their actions in response to the statement (Barton & Sutcliffe, 2009). As a result, voiced content interrupts ongoing action, creating a pause during which members reassess their circumstances, and possibly select a different course of action. Yet, while HRO theorizing depends on employee voice, it often assumes that voice is unproblematic and unhindered, when in fact, there are compelling hindrances. Indeed, a few studies in HRO contexts have found that voice and expertise are intertwined. Barton and Sutcliffe (2009) found that wildland firefighters often remained silent about safety concerns if they doubted their expertise in comparison to others' expertise. Blatt et al. (2006) found that medical residents, who were no longer medical students but not yet physicians doubted the authority of their liminal role and avoided voicing concerns when they thought suggestions might upset the status quo, or supervisors might reject their recommendations. Therefore, expertise crucially informs voice, yet, expertise is neither entirely cognitive nor role-based, particularly in high-reliability organizations. Instead, expertise about complex environments must be learned physically through corporeal performances that put lessons into play (instantiation), test limits (failure), and call on members to sort through numerous social and environmental cues.

Conclusion

This study explored embodied sensemaking in the context of high-reliability organizing, a type of context characterized by constant danger and changing circumstances that often fluctuate in unpredictable ways. Under conditions in which professionals face constant hazards, it is crucial to better understand how they learn to differentiate between hazards and threats, and particularly, when a "normal" hazardous situation is escalating into a threatening one. This study explored critical incident narratives from 37 firefighters to identify how their sensemaking processes involve physically and socially enacted sensemaking performances that often exposed them to danger and, in doing so, exposed deeper insights about hazards, organizational practices, and safety rules. This study identified three embodied sensemaking performances: instantiating performances involved enactments that translated codified knowledge (e.g., safety rules, procedures) from a written to a lived modality (see also Sauer, 1998); failure performances involved acting out the worst-case scenario, and resulted in the individual rethinking procedures and circumstances they already thought they understood, or that they took for granted; and improvisational performances called on members to interact with others to develop in-the-moment creative action under duress, which often prompted them to notice key aspects of situations that were entirely unknown, unseen, or deemed unimportant before the incident. These findings contribute to our understandings about how professionals in an expertise-driven occupation develop embodied experiences, and suggests how embodied experience contributes to members developing a voice for proposing alternatives and offering dissent.

Discussion Questions

1. What does it mean to learn through experience? Are there some things that can only be learned that way?
2. In what ways do we look to other people to influence how we see and interpret confusing situations?
3. If safety rules and procedures are meant to be both standardized, and as straightforward as possible, why do we sometimes need to physically act out a safety rule in order for it to truly make sense?
4. How does embodied sensemaking contribute to developing or strengthening one's sense of voice; that is, one's ability to express dissent or offer alternative courses of action?

References

Aven, T. (2009). Perspectives on risk in a decision-making context—review and discussion. *Safety Science, 47*(6), 798–806. https://doi.org/10.1016/j.ssci.2008.10.008

Barton, M. A., & Sutcliffe, K. M. (2009). Overcoming dysfunctional momentum: Organizational safety as a social achievement. *Human Relations, 62*, 1327–1356. http://doi.org/10.1177/0018726709334491

Blatt, R., Christianson, M. K., Sutcliffe, K. M., & Rosenthal, M. M. (2006). A sensemaking lens on reliability. *Journal of Organizational Behavior, 27*(9), 897–917. http://doi.org/10.1002/Job.392

Brummans, B. H. J. M., Putnam, L. L., Gray, B., Hanke, R., Lewicki, R. J., & Wiethoff, C. (2008). Making sense of intractable multiparty conflict: A study of framing in four environmental disputes. *Communication Monographs, 75*(1), 25–51. https://doi.org/10.1080/03637750801952735

Cunliffe, A., & Coupland, C. (2012). From hero to villain to hero: Making experience sensible through embodied narrative sensemaking. *Human Relations, 65*(1), 63–88. https://doi.org/10.1177/0018726711424321

Flanagan, J. C. (1954). The critical incident technique. *Psychological Bulletin, 51*(4), 327. www.apa.org/pubs/databases/psycinfo/cit-article.pdf

Gremler, D. D. (2004). The critical incident technique in service research. *Journal of Service Research, 7*(1), 65–89. https://doi.org/10.1177/1094670504266138

Hale, A., & Borys, D. (2013). Working to rule, or working safely? Part 1: A state of the art review. *Safety Science, 55*, 207–221. https://doi.org/10.1016/j.ssci.2012.05.011

Hazard (n.d.). *Merriam-Webster Online.* Retrieved from: www.merriam-webster.com/dictionary/hazard

Hindmarsh, J., & Pilnick, A. (2007). Knowing bodies at work: Embodiment and ephemeral teamwork in anesthesia. *Organization Studies, 28*, 1395–1416. https://doi.org/10.1177/0170840607068258

Hultin, L., & Mähring, M. (2017). How practice makes sense in healthcare operations: Studying sensemaking as performative, material-discursive practice. *Human Relations, 70*(5), 566–593. https://doi.org/10.1177/0018726716661618

Ishak, A. W., & Williams, E. A. (2017). Slides in the tray: How fire crews enable members to borrow experiences. *Small Group Research, 48*(3), 336–364. https://doi.org/10.1177/1046496417697148

Jahn, J. L. S. (2016). Adapting safety rules in a high reliability context: How wildland firefighting workgroups ventriloquize safety rules to understand hazards. *Management Communication Quarterly, 30*(3), 362–389. https://doi.org/10.1177/0893318915623638

Kaufman, H. (1960). *The forest ranger: A study in administrative behavior.* Baltimore, MD: Johns Hopkins University Press.

Krieger, J. L. (2005). Shared mindfulness in cockpit crisis situations: An exploratory analysis. *International Journal of Business Communication, 42*(2), 135–167. https://doi.org/10.1177/0021943605274726

Lewis, A., Hall, T. E., & Black, A. (2011). Career stages in wildland firefighting: Implications for voice in risky situations. *International Journal of Wildland Fire, 20*(1), 115–124. https://doi.org/10.1071/WF09070

Maclean, N. (1992). *Young men and fire.* Chicago, IL: University of Chicago Press.

Maitlis, S., & Christianson, M. (2014). Sensemaking in organizations: Taking stock and moving forward. *The Academy of Management Annals, 8,* 57–125. https://doi.org/10.5465/19416520.2014.873177

Mendonça, S., e Cunha, M. P., Kaivo-oja, J., & Ruff, F. (2004). Wild cards, weak signals and organizational improvisation. *Futures, 36*(2), 201–218. https://doi.org/10.1016/S0016-3287(03)00148-4

Mills, C. (2002). The hidden dimension of blue-collar sensemaking about workplace communication. *The Journal of Business Communication, 39*(3), 288–313. https://doi.org/10.1177/002194360203900301

Miner, A. S., Bassoff, P., & Moorman, C. (2001). Organizational improvisation and learning: A field study. *Administrative Science Quarterly, 46*(2), 304–337. https://doi.org/10.2307/2667089

Reason, J. (1998). Achieving a safe culture: Theory and practice. *Work & Stress, 12*(3), 293–306. https://doi.org/10.1080/02678379808256868

Roberts, K. H., Rousseau, D. M., & La Porte, T. R. (1994). The culture of high reliability: Quantitative and qualitative assessment aboard nuclear-powered aircraft carriers. *The Journal of High Technology Management Research, 5*(1), 141–161. https://doi.org/10.1016/1047-8310(94)90018-3

Rochlin, G. I. (1993). Defining high reliability organizations in practice: A taxonomic prologue. In K. Roberts (Ed.), *New challenges to organization research: High reliability organizations* (pp. 11–32). New York, NY: Macmillan.

Sauer, B. (1998). Embodied knowledge: The textual representation of embodied sensory information in a dynamic and uncertain material environment. *Written Communication, 15*(2), 131–169. https://doi.org/10.1177/0741088398015002001

Sauer, B. (2003). *The rhetoric of risk: Technical documentation in hazardous environments.* Mahwah, NJ: Lawrence Erlbaum Associates, Inc.

Threat (n.d.). *Merriam-Webster Online.* Retrieved from: www.merriam-webster.com/dictionary/threat

Tracy, S. J. (2012). *Qualitative research methods: Collecting evidence, crafting analysis, communicating impact.* Malden, MA: John Wiley & Sons.

van Duijne, F. H., van Aken, D., & Schouten, E. G. (2008). Considerations in developing complete and quantified methods for risk assessment. *Safety Science, 46*(2), 245–254. https://doi.org/10.1016/j.ssci.2007.05.003

Weick, K. E. (1979). *Social psychology of organizing*. Reading, MA: Addison-Wesley.

Weick, K. E. (1993). The collapse of sensemaking in organizations: The Mann Gulch disaster. *Administrative Science Quarterly, 38*(4), 628–652. www.jstor.org/stable/2393339

Weick, K. E. (1995). *Sensemaking in organizations* (Vol. 3). Thousand Oaks, CA: Sage.

Weick, K. E. (2003). Positive organizing and organizational tragedy. In K. S. Cameron, J. E. Dutton, & R. E. Quinn (Eds.), *Positive organizational scholarship: Foundations of a new discipline* (pp. 66–80). San Francisco, CA: Berrett-Koehler.

Weick, K. E. (2004). Vita contemplativa: Mundane poetics: Searching for wisdom in organization studies. *Organization Studies, 25*(4), 653–668. https://doi.org/10.1177/0170840604042408

Weick, K. E. (2006). Faith, evidence, and action: Better guesses in an unknowable world. *Organization Studies, 27*(11), 1723–1736. http://doi.org/10.1177/0170840606068351

Weick, K. E., & Sutcliffe, K. M. (2015). *Managing the unexpected: Sustained performance in a complex world* (3rd ed.). Hoboken, NJ: John Wiley & Sons, Inc.

Whiteman, G., & Cooper, W. H. (2011). Ecological sensemaking. *Academy of Management Journal, 54*(5), 889–911. http://doi.org/10.5465/amj.2008.0843

Wolbers, J., & Boersma, K. (2013). The common operational picture as collective sensemaking. *Journal of Contingencies and Crisis Management, 21*(4), 186–199. https://doi.org/10.1111/1468-5973.12027

Ziegler, J. A. (2007). The story behind an organizational list: A genealogy of wildland firefighters' 10 standard fire orders. *Communication Monographs, 74*(4), 415–442. https://doi.org/10.1080/03637750701716594

3

THE CULTIVATION OF SHARED RESOURCES FOR CRISIS RESPONSE IN MULTITEAM SYSTEMS

Elizabeth A. Williams

In the face of what seems to be an ever-increasing threat of natural disasters, terrorist attacks, and health pandemics, the ability to respond to a crisis has become a major focus for most communities. From small towns and large cities to state and national governments and international agencies, emphasis is being placed on emergency preparedness and how various entities will coordinate in order to respond to crises. In the post-9/11 era, the United States' local, state, and federal governments have pushed to make the various teams within emergency response systems coordinate their activities. As Kapucu and Van Wart (2006) contend, emergency response systems are in a state of change as the public and the government expects more of the systems than ever before. We have seen evidence of this over the past two decades with the formation of district Homeland Security planning committees within each state, the implementation of a National Incident Management System (NIMS), and system-wide training protocols. These initiatives help coordinate the efforts of governmental, nongovernmental, and private sector organizations. Indeed, during any emergency, communities may rely on a wide variety of different organizations and agencies during the response effort. The ad hoc system that comes together during a crisis represents a "network of teams." These teams of teams have been coined multiteam systems (MTSs).

This chapter is interested in how these networks of individual teams—each with their own culture, goals, processes, and organizational structure—prepare to come together during crisis events. Specifically, the chapter answers the question: *How do MTSs cultivate their interdependent relationships and shared resources prior to a crisis in order to facilitate effective communication and response during a crisis?* In answering this question, this chapter explores how interdependence is communicated and experienced by system members and how this interdependence influences the ability of MTSs to respond during times of crisis. To do so, this chapter first explores the literature on MTSs and two theoretical perspectives particularly relevant to this project. The next section provides information about the data used to inform this essay. These data reveal three types of resources developed by interdependent teams and four processes used for developing these resources between teams. The chapter ends with a discussion of how practitioners can use their knowledge of these resources and processes to help ensure coordination between the multiple teams of an emergency response system.

Multiteam Systems

MTSs consist of multiple teams working together to achieve a collective goal while negotiating various organizational and team boundaries (Mathieu, Marks, & Zaccaro, 2001). An oft-cited example of an MTS is a team of first responders, paramedics, and surgeons treating accident victims. This example reveals the high stakes in which MTSs often operate and the need to understand the nature of interdependencies among teams. Thus, MTSs offer a context in which to learn and theorize about system processes, pinpoint threats and challenges to coordination, and examine what should be considered as systems that we rely on to effectively respond to crises.

Indeed, because of the interdependent nature of MTSs, there are several inherent challenges to coordination (Mathieu et al., 2001; Williams, 2016). To begin, while MTSs have a shared overarching goal (i.e., respond to the crisis), the individual teams will also have their own proximal goals, some of which may overlap with the overarching goal, and some of which might be in conflict with the overarching goal. For example, when responding to an accident on the highway, all responders want to ensure that the victims of the accident receive adequate care. An additional goal of the highway patrol is to minimize the possibility of accidents in the traffic backup that ensues after an accident. To achieve that proximal goal, the highway patrol is interested in getting lanes of traffic open as soon as possible. This is often not a salient goal for the medical responders and they may even see this as conflicting with their goal of keeping their team members safe from traffic.

Next, these teams may all exist within the same organization or they may come from a variety of organizations, each with their own culture, operating procedures, and communication protocols (Connaughton, Williams, & Shuffler, 2012). For example, when a wildland fire occurs, the first responders are often municipal fire departments. As a fire grows, teams of firefighters from state and national agencies join the response. Even though multiple teams may be coming from one agency (e.g., teams of hotshot firefighters from the National Forest Service), these teams often come from diverse regions of the country and may have had little previous interaction with one another.

The response team, however, does not end with firefighters. Local police authorities work to coordinate evacuations. Local nonprofits work to provide aid to those displaced. Local utility crews, and, oftentimes, additional crews from outside of the immediate area, provide for infrastructure (e.g., electricity, water, telecommunications) needs and assess infrastructure threats. Indeed, looking at the variety of types of organizations involved in a single response, we can begin to understand how different team/organizational cultures, procedures, and communication protocols will exist within the large team brought together to respond to this crisis. Furthermore, these differences exist not only between different types of organizations, but also within organizations as teams respond from a variety of geographic areas.

Despite these differences, these teams must rely on one another—they are functionally interdependent. Functional interdependence is defined as "a state by which entities have mutual reliance, determination, influence, and shared vested interest in processes they use to accomplish work activities" (Mathieu et al., 2001, p. 293). For example, in the wildland fire scenario above, no one team can do their work without the other teams accomplishing theirs. The team of firefighters relies on local law enforcement to evacuate residents, so they can focus on containing the fire. Likewise, law enforcement relies on the expertise of the firefighters to help them determine what areas should be evacuated. Because of these interdependencies, research on MTSs must look beyond within-team processes, and pay special attention to the *between*-team processes. The focus then, when studying MTSs, is the interaction or interdependence between teams (DeChurch & Matheiu, 2009). These between-team processes allow coordination to occur and subsequently support the achievement of system-level goals.

System Relationships and Resources

Exploring between-team processes leads us to examine the relationships that exist between these various teams and the resources available for navigating the interactions among teams. Indeed, MTSs present a complex web of relationships that require careful and thoughtful coordination. Two particular theories that are useful to guide the examination of these relationships and their consequences are leader–member exchange theory (LMX) and structuration theory.

Leader-Member Exchange Theory

LMX explores how the relationship between a team member and the leader of the organization/team influences that member and their performance (Dansereau, Graen, & Haga, 1975; Graen & Scandura, 1987; Graen & Uhl-Bien, 1995). Scholars have expanded this theory by exploring the various different types of relationships that exist within organizations (e.g., between co-workers [CWX], between leaders [LLX]) (Sherony & Green, 2002; Sluss, Klimchak, & Holmes, 2008; Tangirala, Green, & Ramanujam, 2007). Within a MTS, another type of relationship that exists is between different members of the system—that is relationships across teams.

Understanding LMX and its subsequent iterations, allows us see how the existence of this web of within- and across-team relationships helps cultivate resources that enable an effective system response. To explain, as Fairhurst (2007) notes, LMX theory is interested in how relationships influence our ability to understand both *how* to act or communicate and *why* we should. That is, certain discursive resources (i.e., ways of understanding and explaining) become available through the relationships that are fostered between different system members. This chapter is interested in examining those resources that become available to members from their relationships with others in the system. Drawing from LMX, this chapter argues that multiple communicative exchanges—those between the leader and the member, a member's perception of communicative exchanges between their leader and other leaders, and the exchanges the member has with other system members—create the system prior to it be enacted during an emergency. Furthermore, these exchanges enable the system to respond effectively in the face of crisis.

Structuration Theory

A second theory that helps explain how collaboration and coordination occur within an emergency response system is structuration theory. Structuration theory is a meta-theory in that it can be used to help our understanding of a wide variety of phenomenon and can advance other theories. For example, structuration theory has been used to further our understanding of phenomena ranging from the role of technology in organizational change (DeSanctis & Poole, 1994) to shifts in organizational culture (Whiteley, Price, & Palmer, 2013). Specifically, structuration theory focuses on systems of human practices that are influenced by a structure of rules and resources (Giddens, 1979). For structuration theorists, a rule is "any principle or routine that guides people's actions" and resources are "anything people are able to use in action, whether material (money, tools) or nonmaterial (knowledge, skill)" (Poole & McPhee, 2005, p. 174). Within these systems of human practice, individuals act and interact and in doing so, each action "'produces' the practices of which it is a part and it 'reproduces' the system and its structure, usually in a small way, as changed or stable" (Poole & McPhee, 2005, p. 175). To illustrate, in the aforementioned example of an accident on the highway, if the standard protocol is that law enforcement opens additional lanes of traffic as soon as possible and that is the action they take in an incident, their action is reproducing the system. However, if before they reopen a lane of traffic, they seek the input of medical response personnel, they are still reproducing a system that values opening lanes of traffic but they are slightly changing the system to acknowledge the needs of other teams in the system. In essence, the system both shapes actions/interactions and is shaped by them.

The central premise of this chapter is that when individuals recognize that their team exists within a larger system, certain resources become available to them that allow them to act/interact in specific ways. The goal of this chapter is to first delineate the different types of resources that exist for members of emergency response systems and then explore the processes that cultivate these resources. To do so, this chapter examines a specific emergency response system. In the next section, I describe the emergency response system that was studied and how data was collected.

Method

The System

The MTS for this study was the local emergency response system of a Midwestern county. The main organizations comprising this MTS were the police and fire departments for the two largest cities in the county, as well as the police and fire departments at the local university in the county; the state police post; the county policing organization; multiple volunteer fire departments; the local EMT organization; the county emergency management agency (CEMA); the county health department; and a nonprofit organization charged with providing humanitarian relief. By all accounts, this was a well-functioning system. Throughout all phases of the project, system members expressed an understanding and appreciation that the system of which they were a part was unique in the respect that there were not "turf battles" between agencies, leaders worked together, and the system was able to respond to the needs of the community. For the members of this system, this meant that the system was effective.

Data Collection

Data for this chapter comes from interviews conducted with system members from various organizations. Eight of the organizations listed above participated in the study. These organizations included the police and fire departments from Anytown and Everytown (pseudonyms used to protect the anonymity of the community), the state police post, the university fire department, a local chapter of a disaster relief agency, and CEMA. Specifically, respondent interviews (Lindlof & Taylor, 2002) with members of these organizations explored the resources needed for the effective system operation and how those resources were cultivated. The interviews sought to examine how members experience MTS interdependence, when they rely on system resources, and the role of various types of communication (e.g., communication with their leader) in making the system salient.

Participants for interviews were recruited through random sampling of members of each component team. The sampling frames were membership lists provided by the organization. Following Babbie's (2005) recommendations, the lists were numbered and a random number generator was utilized to select individuals to contact for an interview. In addition, leaders were asked for formal interviews (although many of them had talked to me informally when I was negotiating access). Again, depending on the organization, interviews were either solicited via email, letter, or, in the case of one organization, via voicemail. Leaders in the organizations announced that individuals may be contacted so these invitations were not unexpected. Participants decided the location of the interview.

Forty-five individuals (39.8 percent of those invited to participate) were interviewed for this project. Individuals were not given any incentive for participation. The majority of the participants were male, with just four being female and the average tenure with their organization was just under 13 years. Prior to the beginning of the interview, participants signed an informed consent form. With the participant's permission, interviews were tape-recorded and then transcribed. During interviews, participants were asked a series of questions about how they interacted within the system. Interviewees were encouraged to discuss the interdependencies within the system. They also were asked to describe system-focused communication that they have had with their supervisors and co-workers and to illustrate interactions they have with members of other teams in the system. In total, 1,756 minutes of interview data were recorded. The interviews lasted between 21 and 75 minutes each and averaged 39 minutes long. Transcripts ranged in length from ten double-spaced pages to 29 double-spaced pages. In total, there were 799 pages of transcripts with the average interview being 18 pages long. The data presented in this chapter come from a larger study. Questions from the interview protocol specifically addressing the topics in this study represent about a third of the interview data. However, the examples provided throughout the remainder of the chapter come from all parts of the interviews—not just the questions specifically targeting resources or interdependencies. This underscores the prevalence and importance of the development of shared resources in MTSs.

Data Analysis

Analysis of interview data utilized a grounded theory approach (Strauss & Corbin, 1998), with themes emerging from the data. I initially read all of the transcripts and engaged in open coding. Coding was done at the paragraph level initially as to maintain the context of individuals' responses. Seventy initial codes emerged from the data. From here open codes were sorted into 37 axial codes in a spreadsheet. Examples of axial codes include materiality, information flow, and redundancy in training. The research questions guiding this study were then reviewed again and from there, the axial codes were collapsed into selective codes that resulted in the themes presented below. Specifically, three types of resources (material, relational, and knowledge) and four processes through which MTSs cultivate those resources emerged from the data.

Results

Resources for Crisis Response

Each crisis or emergency is unique and requires different responses. To be prepared for the range of different scenarios an emergency response system may face, specific resources must be cultivated at the system level. While individuals typically operate as a member of their individual team, having an understanding of the larger system and their team's role in that system provides specific resources for them to use in response to the crisis. From the interview data, three types of resources emerged: material, relational, and knowledge. In the following sections, I first delineate each of these resources and offer examples of each. Then I explore four processes within emergency response systems that help cultivate these resources across the MTS.

Material Resources

One of the resources that becomes available to members when they have a larger sense of the system is access to material resources. Within this study, material resources are those things that enable the MTS to accomplish its tasks. As we will see through the data below, examples of this would include specific equipment the MTS needs when responding to emergencies or the funding needed to procure that equipment.

One specific example of this was the formation of the "district task force." The district task force was created by the state's Department of Homeland Security and consists of groupings of counties around the state that come together to plan for emergencies and share resources. Everytown Police Captain Morris succinctly explained the district concept, "The state's initiative is to divide the state into districts and be able to channel grants, funding, and equipment that way rather than piece meal county-by-county." All participants who were involved in the district meetings discussed the utility of the meetings. Specifically, they highlighted how the district concept functions to give organization's greater knowledge of and access to material resources (e.g., equipment, grant money). Anytown Firefighter Lee expounded on how teams gained access to material resources through the formation of districts:

> You know it's been a great opportunity for everybody to actually find out what's available in the district because they didn't have any idea that the

> Anytown Fire Department has a HAZMAT team or that CEMA's got a mass decon[tamination] unit or an incident command vehicle. There is a lot of money floating around out there that people don't even have a clue that it even exists. I mean that's, I guess that's maybe one of the downsides of before all this district thing came up. The stuff was there but the only people who knew about it was the people who had it and the people locally who had seen it.

As we see through this example, communication among the various teams in the emergency response system allows them to share material resources and creates a more cost-effective system.

Relational Resources

Next, individuals gain access to relational resources when they interact within with the system. Relational resources refer to the connections that system members have across teams. It encompasses both knowing who someone is and having an idea of how they may respond when responding to an emergency. Anytown Assistant Fire Chief Young indicated the importance of building relationships prior to emergencies when saying, "You know on the incidents, it's not the time to trade business cards." State Police Trooper Anderson further explained:

> I think it's obviously better when you already do have those [relationships], because you feel comfortable with—obviously the more that you work with somebody the better because you can almost feel like—the same thing, like, on a soccer team or like a hockey team, you know that this guy always does this move or whatever, or this is the way he likes to play when he's on the field. So it's obviously better if you have those prior relations.

All participants agreed that having interagency relationships only increases the effectiveness of the system. These relational resources create a sense of familiarity between teams when they are called to act together and allow individuals to have a better sense of how others will react.

Knowledge Resources

Finally, system members become privy to specific knowledge when they understand that their team is part of a larger system. Knowledge resources refer to information that enables system members to make decisions and gain greater situational awareness both before and during an emergency situation.

For instance, Anytown Dispatcher Nelson gave an example of how knowledge is shared when the teams work together:

> Well, the perfect example is this new change we have about letting officers know about these medical alerts, you know that was [our supervisor] who had to fill us in and let us know why they are making the change. You know he has a monthly meeting with all the shift commanders. He always comes back and tells us what goes on in those meetings. It's nice because then we're also aware of what's going on on-the-road with the guys. That is important even though a lot of people don't always think that.

As we see from Dispatcher Nelson's example, the fact that her supervisor is involved in inter-team meetings, gives her access to valuable information. Everytown Fire Chief Edwards explained why shared knowledge is important, "Coming from the same textbook gives an understanding about what roles need to be filled and who's in line to fill them." Just as individuals gain knowledge about the roles individuals will take, they also begin to understand how others will react. State Police Sargent Harris explains why knowledge resources stemming from system familiarity is important:

> I think that when it began these agencies realized that it was going to require [cooperation]. It is a necessity that Joe Smith the deputy knows what Joe Smith the trooper is going to do. So when they get [to an accident scene], we know what we've got to do. Because in so many areas— we do different police work than city officers in [Anytown], than [county police] officers do.

In this quote, we see that even within the same type of teams (e.g., police work), different organizations have different practices and procedures. By knowing how other organizations respond to crisis, there is a better understanding of how a particular team will react and how teams should interact. This knowledge of the system and others in the system serves as an important resource.

Combinations of Resources

To underscore the importance of the variety of resources that are cultivated through an understanding of the larger emergency response system, University Fire Assistant Chief Morgan summarizes the material, relational, and knowledge resources that come from the multiple teams and team members understanding the larger system. He points to the National Incident Management System (NIMS) as a driving force of this understanding and then praises its rewards:

> NIMS has brought us more closer together in the fact that we're becoming more, what's the word, unified. We're responding and using the same

terminologies. We're drilling more and more all the time for the big disasters. And with that is bringing all the agencies closer together and all know what our resources are. And what your capabilities are.

Indeed, if we consider these resources from a structuration standpoint, each time system members utilize the resources (or rules) of the system, they are shaping the system. During the interviews with members of this system, many cited a larger paradigm shift within emergency response. After the events of September 11, 2001, a greater emphasis has been placed on interagency cooperation. Collaboration is now held in high esteem. While the impetus of this shift was traced by many back to the events of September 11, a structuration perspective allows us to better understand how that shift occurs. Each time members have positive interactions with one another, go to joint trainings, or share material resources, they are shifting the rules of the old, more-siloed system and recreating a new system that values interdependencies. These new rules and the sharing of resources helps facilitate system effectiveness.

Pre-Crisis Processes for Developing Interdependence

The establishment of material, relational, and knowledge resources and the development of interdependence between teams does not occur without effort. From participant interviews, the following four processes were identified as cultivating shared resources: establishing redundancy, participating in pre-planned events, employing shared communication technologies, and cultivating positive relationships between leaders. Each of these processes are explored below with specific attention given to how these processes produce shared resources.

Establishing Redundancy in the System

First, members point to joint training, procedures, and the NIMS system as unifying experiences/structures. As participants explained, these mechanisms give all system members a shared understanding of how to respond to emergencies and a common language to use. We can see this as State Police Sergeant Harris talked about the role of similar training:

> But the active shooter training that I'm given, the active shooter training that the [county police] department and the local police departments are given are very much the same. Because it may not be four troopers that show up in an active shooter situation, be it a business or a school. It may be a deputy, a marshal, a couple troopers. So when I show up I know these guys, or these officers. They also know what we're going to do. So with as little amount of discussion as possible, we can conduct operations.

This example shows how joint or equivalent trainings establish redundancy in the system. That is, all members have the same training—they were taught how to respond in the same fashion. The redundancy established through those exercises comes in the form of both knowledge resources and relational resources. As we can see in this quote, the redundancy created when joint training occurs allows for responders to know how others will respond in a crisis situation. Additionally, these joint trainings foster relationships between members of various teams. Almost all participants underscored the need for similar training across teams in order to create redundancy in the system. To explain, the system that comes together in response to an emergency, in this case an active shooter, could be different each time depending on who from each team is on duty and where in the jurisdiction the crisis occurs. Therefore, it is imperative that all members of the emergency response system respond similarly. In those crisis moments, the ability to "conduct operations" is imperative. Establishing redundancy in the system means that no matter who responds they have been trained the same way.

Another example of establishing redundancy in the system is the aforementioned National Incident Management System. NIMS was instituted by the Department of Homeland Security and was created to provide a consistent method of response among emergency response teams around the nation. NIMS, or what many individuals referred to as the incident command system, was lauded in almost all interviews as being the driving force in coordination among teams. University Fire Assistant Chief Morgan expanded on how NIMS works and the benefits of it:

> When it comes into the NIMS system, we have different agencies come in here then they have to they go to your incident commander. So if we go to an outside agency and I'm going to follow what their rules are, I will report to that incident commander and then he will tell me what he needs and then we follow whatever it is that he needs. And then normally you follow within their guidelines... [NIMS] brought every agency together to the same understanding and with the same goals now where before everybody was completely different in how they approached that. Now we're all approaching it the same.

In this example, redundancy is created because the same command system is being used throughout the nation. Across the United States, emergency response systems are unified through NIMS in the way leaders take command of an incident. As University Fire Assistant Chief Morgan explains, NIMS cultivates knowledge resources for system members. The knowledge of who is in charge of an incident is critical in order for a coordinated response to a crisis. Having a national system in place on which all emergency response members are trained is paramount to ensuring an effective response.

Encouraging Participation in Pre-Planned Events

The second way resources are developed in a system is through interactions in non-emergency events. The system in which this research was conducted houses a large university and comprises two cities. Each of these entities hosts events (e.g., sports events, concerts, fairs) that draw thousands of people. A majority of participants referred to these various planned events as opportunities for the system to come together and to practice the principles they would utilize in an emergency situation. Anytown Police Lieutenant Brown expanded on the critical nature of pre-planned events for system effectiveness:

> Because that's what helps prepare us for an actual event if it's going to occur on down the road. By pre-planning events and working together with those we can begin to see how well we work together and what issues come up. Whether it be communication challenges, sharing of information challenges, developing contingency plans. You know you can't wait for an actual event to really work together and be successful and really you know work through something with any kind of success. I mean the pre-planning really is a big—it's just a big training tool just as much as anything to help bring us together to start to work in an environment together.

The pre-planned events in this community offer the system an opportunity to put into practice the protocols and communication systems they would utilize in an emergency—to develop knowledge resources of how the system will work.

Working together through this pre-planned events also develops relational resources. State Police Sergeant Harris shared how success in planning and executing these large-scale community events translates to crisis scenarios: "I'd say for melding five agencies together like we do, I think we do a good job. And I've always told people that. And it spills over. Because the same people that we do that with, we also do crashes." The pre-planned events serve as an occasion for the individuals to build relationships with other system members that are then called upon in crisis situations. Indeed, pre-planned events develop knowledge and relational resources that system members call upon in future emergency scenarios. Furthermore, these events also allow system members to come to an understanding of the material resources that exist within the larger emergency response system.

Employing Shared Communication Technologies

The establishment of shared communication technologies is a critical material resource for emergency response systems. Multiple participants pointed to the importance of having technologies that allowed them to interface with individuals on other teams in the emergency response system. For example, Everytown Police Captain Morris also praised the benefits of a shared radio system and recognized the importance of having an integrated computer system:

> And I think one of the things that has helped us here in [this] county is we're all not only on the same computer network but our same radio network. So we're all, in [this] county, we're all on the same 800-megahertz radio system. And that allows us not only to talk on the verbal side but now we can also on the data side, we can all see and view each other's CAD [computer-aided dispatch] reports, police reports, that type of thing. So that's a big, a big advantage. Before we didn't have a shared computer system. Where we wouldn't know, the left hand wouldn't know what the right hand was doing. You know it would be real hard for police officers to communicate at a scene if we call the [county police] department because they were on a different frequency then us. So now we can go to a common talk group as we call it and we can communicate to each other. Or if we get back here and start typing a report and we notice that we arrested somebody or we have a suspect since we're all in the same computer system we can bring up what we call histories and link these suspects to different cases. Not just within [this city] but all of [this] county and now, with our computer system, we have the ability to go nationwide.

It is clear from this example that the ability to share information—both verbally and through information technology systems—is a critical material resource for the emergency response system. The ability to share information at all times during a crisis—during and after—increases system effectiveness. Indeed these communication technologies are material resources that offer system members ways to connect with each other and serve as an important ways of transferring knowledge.

Cultivating Positive Leader–Leader Relationships

The final process identified as helping develop resources within the larger emergency response system is the development of leader–leader relationships. Because of the hierarchical nature of most emergency response organizations, relationships between leaders are critical. It is through these relationships and the subsequent communication of the strong relationships to organizational members that resources are made available across the system.

The organizations in the emergency response system can be described as hierarchical. There is a clear chain of command and individuals rely on that chain of command when operating within the system. For example, CEMA Director Phillips explained that when he was in law enforcement "there was a very forced rigid rank structure and no matter who, for example, the captain was or who the patrolman was, that patrolman knew he had to defer to the captain. Period." CEMA Volunteer Walker further explains why the chain of command is important:

> You have to have a chief and under that chief you have to have his resources and his resources (people) have their resources which are the workers and everything else. So the accountability. So you're not jeopardizing one or another, or you know. It's just a set organizational flow chart or whatever you want to call it. So there is a chain of command. You don't have rogue people making decisions on their own to make decisions out of the scope of what's going on.

From these descriptions, the respondents describe how the command system is set up to allow decisions to be made and coordination to occur at the higher levels of the organization so that front-line workers can focus on their task.

The command system alone does not create a sense of the system, but combining the strong reliance on a hierarchy with the knowledge that the various leaders in the system are unified in their mission makes the system salient for members. Almost all participants indicated that the leaders in this particular system were very connected. Participants cited monthly meetings of the chiefs or directors as keeping communication lines open. As University Firefighter Green explained:

> I know they have a lot of meetings. A lot of you know administrative meetings and things like that like they will have meetings with the [county police] department where you know there is a [county police chief] and [CEMA] directors and [ambulance] directors and what they do I honestly don't know. I think they do a lot of planning and coordination. You know for big events and things like that. But I know they do a lot of meetings.

As illustrated in this quotation, it is not necessarily imperative to know *what* is being said in meetings among leaders, it is the knowledge that these are occurring that is important. From the leadership perspective, State Police Lieutenant Wilson explains the interactions he has with other leaders and how that influences the men and women who work for him:

> We are really truly blessed here at this district that all of the chiefs, the top folks get along. We have a monthly meeting for that purpose. It's an outstanding opportunity to just kind of come together. And I think as a result of that, there's really no turf battles, everybody gets along, everybody works well kind of for the mission, depending upon what the

situation is. They could call upon us, and we will provide them with whatever resources we can provide, and vice versa. It just kind of depends upon the situation.

In this example, we see that leaders recognize the value of open communication between their teams. The monthly meeting cultivates strong relationships and provides a sense of community and collaboration among the various teams in the system. This translates into a more coordinated system and response.

Temporality of Crisis Planning

When scholars and practitioners think about planning for crisis, they often focus on events that occur before the crisis. However, interviews with participants suggest that effective crisis planning actually occurs before, during, and after a crisis. This is evident throughout the examination of the four processes identified above, because the interactions that occur before the crisis influence action during crisis. In other words, how this system cultivated material, relational, and knowledge resources before an incident directly influenced how the system was able to respond to an emergency. For example, if system members do not have a shared understanding of who is in charge during a crisis event, coordination will be difficult, if not impossible, and response will be negatively affected.

Additionally, when teams debrief after incidents, they often focus on how response can be changed in subsequent crisis—they learn from the interactions they had (Ishak & Williams, 2017). That is, a good after-action review, will identify what resources need to be cultivated before the next emergency. Perhaps additional training needs to be done so team members respond consistently across teams, or possible a clearer chain of command needs to be communicated. Considering this interplay between planning, responding, and reassessing highlights the importance of material, relational, and knowledge resources, not just during crises but also before and after a crisis. Indeed, these processes are ongoing and do not exist only at one point in the crisis but rather resources are being cultivated continuously and the structure of the system is constantly being produced and reproduced through the enactment of these processes.

Implications

This chapter highlights several important implications for members and leaders of emergency response systems, and any multiteam system. First, this research reaffirms that relationships across teams are just as important as within-team relationships. As emergencies become increasingly complex, the teams that respond to those emergencies also become more varied and complex. Focusing on cultivating the relationships between teams and organizations enables access to multiple resources for system members. Leaders of these interdependent teams need to consider how they can provide opportunities for their teams to interact prior to emergencies. Additionally, leaders should recognize the value in communicating to the members of their teams about the efforts occurring at the leader level to ensure coordination between teams.

Second, the three identified types of resources—material, knowledge, and relational—are interdependent. Cultivating one type of resource will help cultivate the others. To illustrate, one of the examples above focused on shared communication technologies. This is a type of material resource. However, having access to this material resource also enables the cultivation of knowledge resources. By having the same communication technologies, teams can observe how other organizations respond to an emergency in real time. By having access to material resources, teams are able to develop knowledge resources. And this interdependent relationship exists between all three types of resources. While developing resources may be seen as time-intensive, leaders in multiteam systems should recognize that the investment in developing one type of resource will also influence the cultivation of other resources.

Next, this chapter highlights the temporal element of the cultivation of resources and the agency that individual teams have in instituting systemic changes. From this research, multiteam system leaders and members are reminded that all actions and interactions they have with one another have the ability to shape and change the system. Perhaps one of the most poignant themes that emerged from this project is that systems are constantly changing. Participants talked extensively about how different the United States' emergency response world is since the events of September 11. And while some of these changes are due to institutionalized changes in the structure (e.g., NIMS), many of the changes occurred because of shifts in how members of various teams communicate with one another. This highlights the ability that each individual and team has to make changes to the larger system through the cultivation of resources.

Finally, this project offers important theoretical contributions related to LMX and structuration theory. First, this study underscores yet another relationship—that of system members—that has the ability to influence individuals' understanding of how and why to act. While LMX initially explained how the relationship between a team member and the leader of the organization/team influences member performance (Dansereau, Graen, & Haga, 1975), more recent work on LMX has shifted focus to other types of relationships within teams (Sherony & Green, 2002; Sluss, Klimchak, & Holmes, 2008; Tangirala, Green, & Ramanujam, 2007). The current work highlights the importance of relationships *beyond* one's immediate work group and organization. Indeed, this study suggests that within an MTS, all relationships have the ability to help with the cultivation of resources. With this emphasis on resources, this chapter adds to the body of literature illustrating how human practices are influenced by structure (Giddens, 1979; Poole & McPhee, 2005). Specifically, this project focuses on the cultivation of structure, in the form of rules and resources, across teams. Underscoring the importance of LMX, central to the creation of structures are the relationships between members of various teams within the MTS. This highlights an important link between LMX and structuration theory.

Limitations and Future Directions

While this project makes some important contributions, there are also limitations and areas that can be built upon. For starters, it is difficult to place boundaries around a system. Indeed, each participant had a slightly different definition of the system. While an attempt was made at gaining access to all the "key players" in the system, some were missing. Similarly, this study was examining the system as a whole rather than examining a response to a specific incident. Future research would benefit from focusing on a specific case and learning from the component teams that were involved in that specific system response. Next, participants in this project highlighted time and time again that this was a high-functioning system. There were few conflicts to which participants could point. While this may indeed be the case, it may also be the case that participants wanted to present their organizations and the system in a positive light. Future researchers should look to specific systems that are in turmoil or have experienced public turmoil in order to determine if the results of this study are consistent when an emergency response system is not high-functioning. Finally, this study relied on the retrospective sensemaking of system members. In other words, it relied on the recollections of participants. Nuance would be added to our understanding of the cultivation and utilization of system resources through an ethnographic study. Indeed, a researcher who was embedded within one of these organizations and was experiencing the emergencies with the team members may be more privy to the actual communication occurring among these system members.

Conclusion

As organizations and communities consider how to plan for the myriad of crises they may face, it is important to emphasize the development of material, knowledge, and relational resources across the multiple teams that will respond during the crisis. These resources are cultivated through various communicative processes. By deliberately engaging in and focusing on these processes, organizational leaders and members will strengthen the ability of their communities to respond in moments of crisis.

Discussion Questions

1. Think about a system of which you are a part. What are the resources that exist within that system? How were those resources cultivated? How do you use those resources?
2. This chapter examines a traditional, official emergency response system. How can the findings of this chapter be translated into other types of multiteam systems? How could you apply the notion of resources and processes to the teams, organizations, or systems detailed in other chapters in this book?
3. A key process that cultivated shared resources in this system was the relationships between leaders. If you were faced with an MTS in which the leaders did not have these strong connections, what actions might you take to fortify the relationships?

References

Babbie, E. (2005). *The basics of social research.* Belmont, CA: Thompson Wadsworth.

Connaughton, S. L., Williams, E. A., & Shuffler, M. L. (2012). Social identity issues in multiteam systems: Considerations for future research. In S. J. Zaccaro, M. A. Marks, & L. A. DeChurch (Eds.), *Multiteam systems: An organization form for dynamic and complex environments* (pp. 109–139). New York, NY: Routledge.

Dansereau, F., Jr., Graen, G., & Haga, W. J. (1975). A vertical dyad linkage approach to leadership within formal organizations: A longitudinal investigation of the role making process. *Organizational Behavior and Human Performance, 13*(1), 46–78. https://doi.org/10.1016/0030-5073(75)90005-7

DeChurch, L. A., & Mathieu, J. E. (2009). Thinking in terms of multiteam systems. In E. Salas, G. F. Goodwin, & C. S. Burke (Eds.), *Team effectiveness in complex organizations: Cross-disciplinary perspectives and approaches* (pp. 267–292). New York, NY: Psychology Press.

DeSanctis, G., & Poole, M. S. (1994). Capturing the complexity in advanced technology use: Adaptive structuration theory. *Organization Science, 5*(2), 121–147. https://doi.org/10.1287/orsc.5.2.121

Fairhurst, G. T. (2007). *Discursive leadership: In conversation with leadership psychology.* Los Angeles, CA: Sage.

Giddens, A. (1979). *Central problems in social theory.* Berkeley, CA: University of California Press.

Graen, G. B., & Scandura, T. A. (1987). Toward a psychology of dyadic organizing. In L. L. Cummings & B. M. Shaw (Eds.), *Research in organizational behavior* (pp. 175–209). Greenwich, CT: JAI Press.

Graen, G. B., & Uhl-Bien, M. (1995). Relationship-based approach to leadership: Development of leader-member exchange (LMX) theory of leadership over 25 years: Applying a multi-level multi-domain perspective. *The Leadership Quarterly, 6*(2), 219–247. https://doi.org/10.1016/1048-9843(95)90036-5

Ishak, A. W., & Williams, E. A. (2017). Slides in the tray: How fire crews enable members to borrow experiences. *Small Group Research, 48*(3), 336–364. https://doi.org/10.1177/1046496417697148

Kapucu, N., & Van Wart, M. (2006). The evolving role of the public sector in managing catastrophic disasters: Lessons learned. *Administration & Society, 38*(3), 279–308. https://doi.org/10.1177/0095399706289718

Lindlof, T. R., & Taylor, B. C. (2002). *Qualitative communication research methods.* Thousand Oaks, CA: Sage.

Mathieu, J. E., Marks, M. A., & Zaccaro, S. J. (2001). Multiteam systems. In N. Anderson, D. S. Ones, H. K. Sinangil, & C. Viswesvarin (Eds.), *Handbook of industrial, work and organizational psychology* (pp. 290–313). Thousand Oaks, CA: Sage.

Poole, M. S., & McPhee, R. D. (2005). Structuration theory. In D. K. Mumby & S. May (Eds.), *Engaging organizational communication theory and research: Multiple perspectives* (pp. 171–196). Thousand Oaks, CA: Sage.

Sherony, K. M., & Green, S. G. (2002). Coworker exchange: Relationships between coworkers, leader-member exchange, and work attitudes. *Journal of Applied Psychology, 87*(3), 542–548. http://dx.doi.org/10.1037/0021-9010.87.3.542

Sluss, D. M., Klimchak, M., & Holmes, J. J. (2008). Perceived organizational support as a mediator between relational exchange and organizational identification. *Journal of Vocational Behavior, 73*(3), 457–464. https://doi.org/10.1016/j.jvb.2008.09.001

Strauss, A., & Corbin, J. (1998). *Basics of qualitative research: Techniques and procedures for developing grounded theory.* Thousand Oaks, CA: Sage.

Tangirala, S., Green, S. G., & Ramanujam, R. (2007). In the shadow of the boss's boss: Effects of supervisors' upward exchange relationships on employees. *Journal of Applied Psychology, 92*(2), 309–320. https://doi.org/10.1037%2F0021-9010.92.2.309

Whiteley, A., Price, C., & Palmer, R. (2013). Corporate culture change: Adaptive culture structuration and negotiated practice. *Journal of Workplace Learning, 25*(7), 476–498. https://doi.org/10.1108/JWL-09-2012-0069

Williams, E. A. (2016). Leading interorganizational collaborations: The importance of multi-dyadic relationships. In T. R. Harrison & E. A. Williams (Eds.), *Organizations, communication, and health* (pp. 330–346). New York, NY: Routledge.

SECTION II

How Individuals Seek, Share, and Get Messages

Section II of this book focuses on individuals' actions and how they seek, share, and receive crisis messages. This section begins with a focus on risk-information seeking around safety information at work. Here, Ford discusses the foundational concept of uncertainty and how many of safety messages sent and received revolve around managing uncertainty. Next, Cacciatore, Kim, and Danzy examine another common reality for organizations: when their reputations are threatened by crises played out through public social media platforms. These scholars examine two different crises experienced by United Airlines, and their social-listening analysis shows how the public's sentiment changed and fueled the crisis.

While discussions on new media are spread throughout this book, this section examines a form of new media that is being used to push or send messages. The photo in Figure 0.3 is a screenshot of a wireless emergency alert (WEA) message the first time a U.S. president sent this type of message to the public.

In Chapter 6, Bean and Madden discuss these WEAs, and the efforts being made to help them become productive emergency message platforms. This section ends with a comprehensive review of literature on evacuations, a common reason that WEAs may be sent in the future. Transportation and civil engineers Rambha, Jafari, and Boyles, share literature on transportation, traffic, shelters, and the various computer models and simulations used to predict human behaviors.

FIGURE 0.3 Screenshot of a WEA broadcast in the U.S. in 2018

4

IDENTIFYING COMMUNICATIVE PROCESSES INFLUENCING RISK-INFORMATION SEEKING AT WORK

A Research Agenda

Jessica L. Ford

"The best high reliability organizations know that they have not experienced all of the ways that their system can fail" (Weick & Sutcliffe, 2007, p. 3). High-reliability organizations are harbingers of daunting demands, risks, and accidents, which sets them apart from other types of organizations (Weick & Sutcliffe, 2007). It is no surprise then, that construction workers, firefighters, miners, and oil workers regularly incur the highest number of on-the-job injuries and fatalities (U.S. Bureau of Labor Statistics, 2016). Recognizing that workplace injuries and fatalities are indications that there has been a "failure" in organizational communication (Real, 2008), it is surprising that research on this topic is scant (see Real, 2008, 2010; Zoller, 2003 for notable exceptions). In fact, Barling, Loughlin, and Kelloway (2002, p. 488) note, "less than 1% of organizational research published in top journals has focused on occupational safety, a situation that has not changed for two decades." Considering that 2.9 million nonfatal occupational injuries occurred in 2016 alone (U.S. Bureau of Labor Statistics, 2017), it is shocking that such little fervor for research on occupational health and safety exists within organizational communication scholarship.

Because workplace injuries affect health, but happen at work, approaching this topic requires interdisciplinary contributions. In the only other appeal for communication research on this subject, Real (2010) recognizes the value of interdisciplinary work, calling specifically for "health-related organizational communication" research (p. 457). This chapter further develops Real's (2010) call for interdisciplinary research on occupational safety by presenting a cursory review of organizational, crisis, and health communication scholarship on information seeking. All three domains of research help identify different communicative aspects of information seeking, which I then use to create a research agenda on employee risk-information seeking. For instance, while organizational communication research on information seeking focuses primarily on socialization, performance evaluations, and organizational change (see Morrison, 2002 for a review), the present research highlights the need to study these three areas as communicatively constructed processes that enhance our understanding of risk information seeking. Thus, learning from other approaches to information seeking enriches future studies on risk-information seeking in the workplace. Seeing as communication research on risk-information seeking helps pinpoint possible obstructions to workplace safety, this area of research demands more attention.

There are, however, exceptions to this general pattern of oversight for organizational risk-related research. Zoller (2003) shows how discursive practices in the workplace foster employee consent to health hazards and silence injured employees. Often, employees do not report work-related injuries because of tensions between occupational identity, organizational norms, and regulatory organizational structures (Zoller, 2003). Scott and Trethewey's (2008) examination of firefighters also uses a discursive approach to risk navigation at work, finding that occupational identity constrains risk management practices. Despite physical hazards at work, these firefighters relied on group-level discourse, rather than regulatory processes to substantiate risk (Scott & Trethewey, 2008). Real (2008) also contributes to this literature by demonstrating that responses to workplace risks are a function of an individual's safety and efficacy beliefs, and later calls for more communication research on occupational safety (Real, 2010).

These examples of communication research on occupational risk do not represent a disciplinary trend, nor do they completely uncover the communicative processes influencing risk-information seeking. Although there are journals devoted to safety research (e.g., *Safety Science, Journal of Safety Research, Accident Analysis and Prevention*), these outlets focus on the antecedents for safety compliance rather than the communicative processes influencing risk-information seeking (see Cooper & Phillips, 2004; Paul & Maiti, 2007; Vredenburgh, 2002). Research on safety and risk does, however, highlight safety culture as a significant predictor of safety behaviors (Cooper & Phillips, 2004; DeJoy, 2005; Fernández-Muñiz, Montes-Peón, & Vázquez-Ordás, 2007; Vinodkumar & Bhasi, 2009). Because communication research is ideally suited to study aspects of organizational culture, such as safety culture, I see these findings as a reference point for future research from communication scholars on risk-information seeking at work. Although I acknowledge the value safety research contributes to our understanding of workplace behaviors, I petition for more attention to be paid to the communicative mechanisms driving risk-information seeking, rather than compliance. Thus, this chapter uses communication literature on information seeking to create a research agenda for future organizational communication scholarship.

In the sections that follow, I first explain why research on occupational risk-information seeking necessitates an uncertainty management perspective. Subsequently, I review organizational communication literature on information seeking, noting how each area of research in this field points to a set of important contexts and communicative dynamics for future studies on risk-information seeking. Next, I review crisis and health communication research on information seeking, since these domains embody aspects of organizational life within industries characterized by risks and hazards. As such, I adopt and extend the findings from crisis and health communication research on information seeking to build a research agenda on employee risk-information seeking. Finally, I summarize the lessons organizational, crisis, and health communication literature imparts on future risk-information seeking scholarship.

Risk-Information Seeking as Uncertainty Management

Uncertainty is perhaps the most common experience among organizational workers. Morrison (2002, p. 229) states "organizations are institutions characterized by ambiguity, change, and uncertainty." These three features drive information-seeking behaviors, but none spur more research than uncertainty. Literature on uncertainty management within organizational communication covers a range of topics from socialization (Kramer, 2010; Kramer, Dougherty, & Pierce, 2004) and performance evaluations (Ashford & Cummings, 1983), to workplace relationships (Kassing, 2000) and organizational change (Lewis, 1999). Yet, one aspect is clear throughout this research: information seeking is central to the management of uncertainty (Morrison, 2002).

Uncertainty exists when an individual perceives information to be unavailable, inaccessible or inconsistent (Brashers, 2001). Whereas previous scholarship promotes uncertainty reduction as the compulsory response to uncertainty (Berger & Calabrese, 1975), more recent research recognizes that uncertainty reduction is only one of numerous responses to situations where information is lacking (Brashers, 2001). In addition, uncertainty reduction may not be possible in all situations, thus increasing the need for an alternative perspective. To better explain all behavior under uncertain organizational circumstances, I use uncertainty management theory as a guiding lens for research on employee risk-information seeking (Brashers, 2001).

Brashers best describes the scope of uncertainty management when he states, "Communication in uncertainty management follows from appraisals and emotional responses; it encompasses managing uncertainty that is challenging; managing uncertainty that is essential for maintaining hope, learning to live with chronic uncertainty, and managing information problems" (Brashers, 2001, p. 482). Considering that each of the situations Brashers describes is present for employees who work in the midst of hazards (e.g., firefighters, construction workers, miners), it is appropriate to use uncertainty management as the guiding theoretical framework for future research in this area. The strength of uncertainty management theory is that it allows researchers to consider the multilayered complexity between risk assessment and information seeking. Since information influences the assessment of risks in the workplace (Griffin, Neuwirth, Dunwoody, & Giese, 2004), researchers need to consider how personal, relational, and contextual features challenge uncertainty management.

Uncertainty management theory is a dominant framework in organizational, crisis, and health research because it brings attention to the experience of uncertainty and the creation of meaning in a variety of contexts. Since the present research on risk-information seeking contains organizational, crisis, and health-related aspects, Brashers (2001) uncertainty management theory aptly frames this research agenda. The following sections review the literature on information seeking from these three areas of communication research noting how each body of work helps to direct future inquiry on the communicative processes influencing employee risk-information seeking.

An Interdisciplinary Approach to Risk-Information Seeking Research

Reviewing the literature on risk-information seeking alone only provides a narrow glimpse into the communicative processes influencing this behavior. A broader consideration of information seeking, however, exposes the value of adopting and extending other research approaches to study employee risk-information seeking. Specifically, organizational communication research on information seeking highlights socialization, performance appraisals, and organizational change as important communicative processes for future studies on risk-information seeking to consider (e.g., Kramer et al., 2004; Morrison, 2002; Myers et al., 2015). Crisis communication research on information seeking points future research on employee risk-information seeking to consider how individuals select sources of information and discursively construct risk resilience (Chewning, Lai, & Doerfel, 2013; Stephens, Barrett, & Mahometa, 2013). And health communication research reminds us of the role social comparisons play in predicting risk-information seeking (Bigman, 2014). In the following sections, I use the present literature on information seeking from organizational, crisis, and health communication scholars to create a research agenda on employee risk-information seeking.

Organizational Information Seeking

Employees face numerous situations where information seeking is an appropriate response to uncertainty. Job interviews, newcomer interactions, task-related procedures, change management, and performance appraisals all induce uncertainty for employees (Morrison, 2002). Although navigating risks at work also causes uncertainty, organizational communication research examining risk and health information seeking is scant (see Real, 2010; Zoller 2003 for notable exceptions). Instead, the majority of research on information seeking by organizational scholars looks at socialization, performance feedback, and change implementation (Morrison, 2002). Yet, each of these three areas helps direct future scholarship on risk-information seeking.

Socialization

Socialization refers to the process of joining and assimilating into an organization (see Kramer & Miller, 2014 for a review). Communication during socialization structures newcomers' behaviors and their identity at work (Poole, 2011), which is important for scholars to explore when addressing the ways employees interact with workplace hazards. The socialization process can help explain why Scott and Trethewey (2008) found both new and veteran firefighters not wearing their protective breathing masks when cleaning burn sites. Since veteran firefighters would tease new members of the squad who wore their breathing masks, newcomers had to decide to either engage in this protective behavior or fit in with the rest of the squad. Recognizing that the social pressure to conform can put organizational members in danger, more research must address how communication during socialization impacts occupational injuries and fatalities. In particular, how do newcomers acquire risk-related information from more experienced organizational members during socialization?

Research on socialization also looks at unmet expectations. In general, the research on unmet expectations refers to the discrepancy between expectations created during the recruitment process and the reality of the job (Jablin, 2001). Extending the research on unmet expectations, scholars need to consider how safe new employees expect to be at work and whether safety expectations predict risk-information seeking. Bearing in mind that expectations create "blind spots" and threaten our ability to recognize hazards (Weick & Sutcliffe, 2007, p. 23), it is appropriate to look at the formation of safety expectations during socialization. Perhaps those who begin work in high-reliability organizations are more comfortable with hazards, and as a result, overlook the importance of information on safety, hazards, and risks.

Performance Feedback

Seeking information within an organization comes at a cost (Morrison, 2002; Real, 2010). Research on employee performance feedback seeking reminds scholars interested in the ways workers acquire risk information that information seeking is an intentional and calculated process (Morrison, 2002; Myers et al., 2015). Research demonstrates how a combination of high uncertainty and low uncertainty tolerance motivates more frequent information seeking (Ashford & Cummings, 1985; Bennett, Herold, & Ashford, 1990). In addition to evaluating uncertainty tolerance, employees also consider how threatening feedback seeking is to their public image and ego (Morrison & Cummings, 1992). Bearing in mind that organizational norms help appraise the costs and benefits associated with information seeking, future research needs to consider the relative influence of organizational norms on risk information seeking.

Organizational Change

Research on organizational change indicates that these punctuated events create uncertainty and, thus, information needs (e.g., Kramer et al., 2004; Lewis, Schmisseur, Stephens, & Weir, 2006; Susskind, 2007). Again, the desire to make sense of new situations motivates information seeking (Morrison, 2002). In their study on organizational downsizing, Casey, Miller, and Johnson (1997) found employees who survived massive cuts were more insecure about their job stability and used information-seeking strategies to manage their uncertainty. Similarly, incidents that result in workplace injuries may produce information needs not only for those that are hurt, but those that "survive" these events without getting hurt. Future research needs to consider how workplace injuries create organizational changes in employees' information needs as well as seeking behaviors.

Casey and colleagues found that when employees experience information deprivation during organizational change, they are less likely to use overt information seeking strategies (Casey et al., 1997). This is an important finding for the present research, considering that not all organizations provide adequate safety information to their employees (Real, 2008). As a result, employees may be in an information void; where they are uncertain about how to navigate workplace hazards, but unable to look for risk information overtly. Thus, scholars need to explore how employees experiencing structurally imposed information voids covertly seek risk information (see Ford, Stephens, & Ford, 2014).

Crisis Communication Research

Organizational crisis literature focuses primarily on large-scale events that threaten organizational stability. For instance, the BP oil spill (Watson, 2015), the 2007 toxic pet food catastrophe (Stephens & Malone, 2010), and the Toyota recall (Choi & Chung, 2012) are all examples of crises studied by communication scholars. Because crisis research focuses on events that affect organizations as a whole, there is a noticeable gap in crisis research on events that cause occupational injuries. Although injuries at work may not disable an organization to the extent of an event like the BP oil spill, the fact that there were 2.9 million reported cases of injury-related absenteeism in 2016 (U.S. Bureau of Labor Statistics, 2017) is evidence of a widespread problem. If a crisis is measured by its effect size, then certainly workplace injuries warrant a crisis lens.

The following section uses crisis communication research on information sources and post-crisis resilience to help guide future scholarship on workplace risk-information seeking. Overall, these two areas of research demonstrate how communication during change both reifies and recreates ways of communicating. Extending these findings to the present research on risk-information seeking provokes new, unexplored directions for organizational communication scholars.

Sources of Information

By definition a crisis creates uncertainty, which individuals reduce or manage by seeking information (Sellnow, Seeger, & Ulmer, 2002). The "sufficiency principle" is often an implicit assumption within crisis communication research, which states that individuals continue to seek information until they no longer feel a threatening level of uncertainty (Sommerfeldt, 2015). It is no surprise, then, that much of the crisis communication research looks at the sources of information people select during and after an emergency (e.g., Austin, Liu, & Jin, 2012; Liu, Austin, & Jin, 2011; Perry, 2007; Procopio & Procopio, 2007; Spence, Lachlan, & Burke, 2011). To a large extent, crisis communication research examines why a particular medium is chosen and what conditions predict medium selection. Other scholars, however, recognize that individuals use multiple sources to find and verify information during and after a crisis (Seeger, Sellnow, & Ulmer, 2003; Spence et al., 2005; Stephens et al., 2013). Research on source selection indicates that individuals choose sources based on their availability and perceived credibility (Stephens et al., 2013; Westerman, Spence, & Van Der Heide, 2014). Likewise, employees looking for information about workplace risks may utilize a number of sources to retrieve information. Scholars who study risk-information seeking in organizations need to consider how employees use a combination of friends, family members, managers, as well as current and previous co-workers to provide information about workplace risks.

Furthermore, research on the predictors of information seeking show that there are demographic differences in the types of sources individuals primarily use for information (e.g., Sommerfeldt, 2015; Spence, Lachlan, & Griffin, 2007; Taylor-Clark, Viswanath, & Blendon, 2010). For example, in their study on information seeking after the 9/11 attacks, Spence and colleagues find that females used older forms of media (i.e., television and radio) to find information, whereas males relied more heavily on the Internet (Spence et al., 2006). Similarly, Sommerfeldt's (2015) study on the 2010 earthquake in Haiti illuminates an education-based media usage divide, where those with more education use "elite" information repertoires (i.e., Internet and text messages), and lower levels of education use "traditional" information repertoires (i.e., word of mouth, television, and church; Sommerfeldt, 2015). Although there are fewer information sources in organizations than those available to the public during a crisis, the previous literature brings attention to certain aspects of risk-information seeking that are overlooked in the present literature.

This past research suggests several areas for future research. First, researchers need to consider how employees use a combination of sources to locate information about workplace hazards and risks. Second, demographic features may predict certain patterns of source selection within organizations. For instance, researchers need to look at the way employee tenure affects risk-information seeking. Beyond calling attention to source selection for information, crisis communication literature on resilience also helps advance research on risk-information seeking in organizations.

Crisis Resilience

"Unexpected events often audit our resilience" (Weick & Sutcliffe, 2007, p. 1). They uncover weaknesses in our ability to recover, adapt, and return to "normal." Resilience refers to the "ability to repair old practices and develop new practices when the old ones are no longer possible" (Mark, Al-Ani, & Semaan, 2009, p. 690). From a communication perspective, recovery efforts are both material and discursive (Zoller, 2003). For example, natural disasters, such as Hurricane Katrina, required business in New Orleans to both physically rebuild and communicatively restructure business practices (Chewning et al., 2013). Similarly, resilience after workplace injuries involves restoring physical and communicative disruptions. Keeping in mind that many injuries occur because of misused or malfunctioning machinery (Driscoll, Harrison, Bradley, Newson, 2008; Trethewy & Atkinson, 2003), workplace hazards are often inherently material (Zoller, 2003), while the injuries themselves are physical. Yet it is through communication that individuals and organizations reconstruct work processes and social relationships after an occupational injury (Zoller, 2003). Thus, communication is the precursor to resilience, not material restoration.

In line with the conceptualization that resilience is a communicative competency, Buzzanell (2010) identifies five processes that generate resilience: (a) crafting normalcy, (b) affirming identity anchors, (c) using communication networks, (d) exercising alternative methods of achieving goals, and (e) restraining negative emotions. I argue that these five processes are entry points for future communication scholars to study resilience after occupational injuries at either an individual or organizational level. Due to space limitations, however, I only elaborate on how future organizational research needs to consider the ways employees communicatively craft normalcy and navigate their social network to normalize post-injury work life.

First, normalcy is crafted in dialogue with others (Buzzanell, 2010). Thus, the "return to normal" after an occupational injury is as much a physical outcome as it is a communication-based process. Research on organizational reentry after an injury validates the importance of communication between the injured person and their co-workers (Mansfield, Stergiou-Kita, Kirsh, & Colantonio, 2014). Employees returning to work after an injury often experience humiliation and shame because their coworkers and supervisors question the legitimacy of their injury (Dembe, 2001; Kirsh, Slack, & King, 2012). Recognizing that the social costs of injury, such as shame and stigmatization, are expressed through communicative acts, it is evident that on-the-job injuries disrupt workplace communication. Thus, scholarship on organizational health and safety needs to look at resilience through "normalcy discourse and performance" (Buzzanell, 2010, p. 3). Specifically, research should consider whether risk-information seeking after an injury predicts normalcy discourse and performance.

Second, resilient individuals maintain their communication networks and nurture new connections (Buzzanell, 2010). In essence, the resources available to an individual or an organization during a crisis, defined as social capital, enhance their capacity to restore normal operations (Doerfel, Lai, & Chewning, 2010). Doerfel and colleagues (2010) use their research on organizations affected by Hurricane Katrina to illustrate the importance of pre-crisis relationships on post-crisis resilience. Although new connections can be made after a crisis, organizations typically rely on pre-existing relationships to share information and resources post-crisis (Chewning et al., 2013). As a result, organizations with a variety of weak and strong ties (Granovetter, 1973) recover quicker than isolated organizations or ones with few ties (Doerfel et al., 2010).

These findings are helpful for scholars looking to test if resilience among injured employees rests on the same principles of Granovetter's (1973) weak and strong ties. Do employees who have limited relationships at work experience a longer road to resilience? For instance, employees who work alone with dangerous machinery may not be connected to others in their organization the same way that a veteran firefighter is connected to his or her squad. How, then, do individuals with limited communication networks in an organization seek resources, support, and information following an on-the-job injury? Does risk-information seeking look different for previously injured employees with highly enmeshed communication networks than for disconnected workers? More research needs to consider how injured employees activate their communication network through a combination of human interaction and information communication technologies (ICTs). This research is especially important for safety practitioners in charge of fostering a resilient workforce within high-reliability organizations, which experience more work-related injuries and fatalities (Weick & Sutcliffe, 2007).

Health Communication Contributions

The study of information seeking and risk management embodies much of health communication literature. Like organizational and crisis communication, health scholars also recognize uncertainty as a dominant driving force for behavioral change (see Brashers, 2001; Brashers, Goldsmith, & Hsieh, 2002). Health communication research helps direct organizational inquiry on risk-information seeking in three ways. First, the information-seeking models developed by health communication scholars offer guiding frameworks for research on employees' risk- information management (see Table 4.1; Afifi & Weiner, 2004; Ajzen, 1991; Berger & Calabrese, 1975; Brashers, 2001; Freimuth, Stein, & Kean, 1989; Griffin, Dunwoody, & Neuwirth, 1999; Johnson, 1997; Kahlor, 2010; Katz, Blumler, & Gurevitch, 1973; Sunnafrank, 1986; Witte, 1992). Second, health scholars call attention to other forms of information management besides active seeking (e.g., Afifi & Afifi, 2009; Afifi & Weiner, 2004; Barbour, Rintamaki, Ramsey, & Brashers, 2012; Brashers, 2001; Brashers, Neidig, & Goldsmith, 2004; Case, Andrews, Johnson, & Allard, 2005). Third, health communication scholars recognize the important role social comparisons play in promoting protective behaviors (Bigman, 2014), which prompts organizational scholars to consider this dynamic within the workplace. This section highlights the value of health communication literature in guiding future research on employee risk-information seeking about workplace hazards.

Modeling Information Seeking

Recognizing the importance of information on health outcomes, health communication scholars continually develop new models to capture the antecedents and outcomes of information seeking (see Table 4.1). Although variations among the information seeking models indicates a lack of conceptual clarity (Lambert & Loiselle, 2007), these models continue to drive empirical research on health. Depending on what aspect the researcher is interested in studying—source of information, antecedents to seeking, channel used to acquire information, outcomes of seeking, or a specific context—there exists a corresponding model. For instance, the Comprehensive Model of Information Seeking (CMIS) developed out of a need to understand how people select a channel for seeking cancer-related information (Johnson, 1997).

Given that each model has a unique treatment of information seeking, it is appropriate to look at risk-information seeking as distinctive from other types of health information seeking. Kahor's (2010) Planned Risk-Information Seeking Model (PRISM) provides a superb foundation for research on risk-information seeking in the workplace (see Figure 4.1). Kahlor defines health risk-information seeking as "the effort expended to locate information about risks to one's personal health" (2010, p. 346). The PRISM rests on the assumption that risk-information

TABLE 4.1 Review of information management models used in health communication research

Model name	Author(s)	Goal of model or theory	Variables or factors of interest	Strengths	Weaknesses
Comprehensive Model of Information Seeking (CMIS)	Johnson (1997)	Predict how individuals seek information through specific information carriers.	Demographics, experience, salience, beliefs, and medium utility/characteristics.	Focuses on the channel the information seeker selects.	Support of model differs depending on the context. Factors too broad such that it is a loosely causal. No feedback loop.
Theory of Planned Behavior (TPB)	Ajzen (1991)	Explains the causal dynamics responsible for human behavior.	Attitude toward behavior, subjective norm, perceived control, intention, and behavior.	Accounts for actions that are less voluntary. Strong predictive power in studies.	Does not provide the sequence or the relative importance of the variables on human behavior.
Risk Information Seeking and Processing (RISP)	Griffin, Dunwoody, & Neuwirth (1999)	Explains the variance in information seeking when risk is present.	Risk information gap, informational subjective norms, channel beliefs, self-efficacy.	Evidence that social pressure encourages people otherwise uninterested in risk information to seek and process this content.	Treats knowledge as an exogenous variable, limiting its influence on the information–seeking process. Does not account for attitudes toward seeking.
Uses and Gratifications	Katz, Blumler, & Gurevitch (1973)	Describes why and how certain media are selected to meet information goals.	Needs: cognitive, affective, personal, social and stress relieving.	Valued theory among computer–mediated communication researchers and helps explain the selection of certain media over others.	Lacks single general theory. Broad concepts hard to operationalize. Assumes individuals always desire more information. Views humans as active seekers of information.

(Continued)

Model name	Author(s)	Goal of model or theory	Variables or factors of interest	Strengths	Weaknesses
Health Information Acquisition Model (HIAM)	Freimuth, Stein, & Kean (1989)	Explains active information seeking. Depicts a process that is either continued or terminated at each step.	Stimulus, information goals, cost–benefit analysis, search behavior, information evaluation, and decision.	Recognizes iterative and goal driven nature of information seeking. Considers intra/interpersonal sources as well as mass media.	Does not account for other information acquisition strategies other than active seeking. Does not look at personal or contextual factors affecting information seeking.
Uncertainty Reduction Theory (URT)	Berger & Calabrese (1975)	Predict and explain first-time interactions between strangers.	Verbal communication, self-disclosure, nonverbal warmth, reciprocity, liking, information seeking, and similarity.	Extensively applied in a variety of contexts, including organization socialization and intercultural interactions.	Mixed support in empirical studies. Assumes uncertainty is negative and should always be reduced. Believes human behavior follows universal laws.
Extended Parallel Process Model (EPPM)	Witte (1992)	Prescribes the use of fear as a motivator of protective responses.	Message, threat appraisal, efficacy, and response (constructive or defensive).	Proven ability to predict which fear-based health campaigns will be effective.	Assumes audiences have no prior awareness to the threat prior to message exposure. Also assumes threat appraisals take place constantly instead of sequentially or at a certain time.
Uncertainty Management Theory (UMT)	Brashers (2001)	Describe how people manage their uncertainty and others' uncertainty.	Uncertainty appraisal, seeking/avoiding information, adapting to uncertainty, social support.	Allows for the consideration of other ways to manage uncertainty besides reduction.	Does not predict when individuals will choose certain uncertainty management strategies over others.

Model name	Author(s)	Goal of model or theory	Variables or factors of interest	Strengths	Weaknesses
Predicted Outcome Value (POV)	Sunnafrank (1986)	Explains behavior during initial interactions, where uncertainty is present. Provides different explanation than URT.	Goals, predicted outcome value, and uncertainty.	Well supported findings that outcome maximization, not uncertainty reduction, is the primary motivator of communication behavior in initial interactions.	Assumes individuals actively assess the outcome value of all interactions. Does not account for previous interactions. Used more as a general theory than as a model for research.
Theory of Motivated Information Management (TMIM)	Afifi & Weiner (2004)	Explains and predicts the decision-making process involved in reduction anxiety through information management.	Uncertainty discrepancy, anxiety, outcomes, efficacy, and information management strategy.	Considers multiple types of information management strategy (avoiding, blunting, coping, seeking, etc.). Includes the role of the information provider.	Capturing the role of the information provider in the model is confusing. Most studies only test the information seeker portion of the model. Theory bound to inter-personal contexts.
Planned Risk Information Seeking Model (PRISM)	Kahlor (2010)	Explain and predict the individual-level variables that influence health risk-information seeking.	Risk perception, affective response to risk, knowledge discrepancy, attitude toward seeking, perceived seeking control, seeking-related subjective norms, and seeking intent.	Draws relationships among variables from other well-studied information management models. Maps the complexity behind risk information seeking specifically.	Treats information seeking as a deliberate behavior which discredits other forms of information management. Only measures seeking intent. No defined entry point into the model.

seeking is a planned behavior largely influenced by individual-level variables (Kahlor, 2010). The PRISM posits that there exists a web of relationships among (a) seeking-related norms, (b) perceived seeking control, (c) attitude toward seeking, (d) risk perception, (e) affective risk response, and (f) risk knowledge discrepancy on an individual's decision to seek risk information. Because the PRISM is designed to capture risk-information seeking, and adopts the most relevant concepts from previous information management models, I urge future scholars to test this model within an organizational setting.

The PRISM captures the complexity of information seeking within an organization by recognizing the role of norms and perceived seeking control on an individual's desire to look for information about health-related risks. Whereas previous scholarship acknowledges safety culture as a predictor of safety-related behaviors (see Vredenburgh, 2002 for a review), research exploring the relationship between normative organizational behaviors and an individual's perceived seeking control on risk-information seeking is limited. One exception is Ford and Stephens' (2018) research on employee risk responsiveness. Ford and Stephens (2018) expand on many of the core relationships found in the PRISM to create and test an integrated model of risk responsiveness in high-reliability organizations. The results of their model testing indicate that individual factors—tolerance for ambiguity, risk perception, and attitude toward risk-information seeking—in addition to perceptions of organizational processes—information seeking norms and seeking control—impact employees' risk responsiveness (Ford & Stephens, 2018). In light of these findings, future research should consider not only the communicative construction of risk-related norms, but also the structural features that constrain risk-information seeking in an organization.

Alternative Responses to Uncertainty

In addition to contributing valuable information-seeking models, health communication scholars recognize that seeking is only one of many outcomes to uncertainty (Brashers, 2001). Besides seeking, health scholarship provides evidence for information management strategies such as avoiding (Afifi & Afifi, 2009; Ramanadhan & Viswanath, 2006), blunting (Case et al., 2005; Folkman & Lazarus, 1980; Miller, 1987), and coping (Hines, 2001; Hines, Babrow, Badzek, & Moss, 2001) when confronted with uncertainty. Since the desire to reduce uncertainty is not ubiquitous, Brashers' (2001) uncertainty management theory better explains behavior during uncertainty than Berger and Calabrese's (1975) uncertainty reduction theory. The recognition of other information management strategies prompts organizational scholars interested in studying how employees navigate risks in the workplace to consider other outcomes of uncertainty. For instance, what are the predictors of coping with risks at work versus active risk-information avoidance? To what extent does an employee's risk perception and perceived risk knowledge insufficiency predict his or her information management strategy? Once again, drawing on

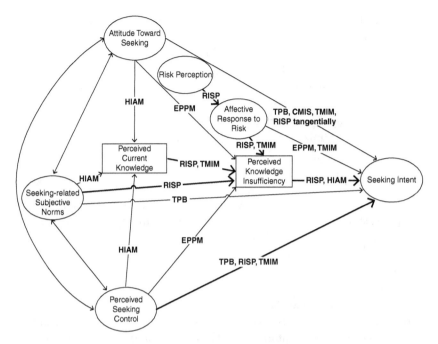

FIGURE 4.1   The Planned Risk-Information Seeking Model (PRISM). This model is taken from Kahlor's (2010) conceptualization of the PRISM. The acronyms along the paths indicate the models from which these relationships were originally taken.

Permission granted by Taylor & Francis to re-print this figure.

interdisciplinary research on uncertainty management helps generate new directions for research on organizational health and safety.

The Role of Social Comparisons

Health literature on risk-related social comparisons guides organizational scholars to also address the influence of social comparisons on risk information seeking. Disparity research rests on the premise that differences in certain features (e.g., race, class, socioeconomic status) impact the likelihood of future behaviors (CDC, 2011). As a result, one group assumes more risk for negative health outcomes than another group (Bigman, 2014). Recent research calls attention to the influence of social comparisons on framing risk perceptions. For instance, in his study on racial social comparisons and risk perceptions, Bigman (2014) found that social comparison lowered perceived health risks for less at-risk groups, but did not raise risk perceptions for more at-risk groups. This finding helps explain reverse health disparities, where low-risk groups dismiss their susceptibility to certain health conditions, such as sexually transmitted infections (Wiehe, Rosenman, Wang, & Fortenberry, 2010), and end up with worse survival rates due to late diagnosis. Extending this finding to the present area of study demonstrates the need to consider how social comparisons at work frame risk perceptions. Perhaps certain personal features, such as muscular build or fitness level, are used as a measuring stick for employees to calculate their risk to occupational injuries in certain industries. As a result, organizational workers who perceive that their risk for occupational injury is low may not seek risk information. Using the findings from health scholars on social comparisons provides another entry point for organizational inquiry on risk-information seeking.

Practical Implications

Because of the limited literature on employee risk-information seeking, we must look at how other areas of communication research approach information seeking in general. Organizational, crisis, and health communication literature highlights various communicative processes influencing information seeking, which may also impact employee risk-information seeking. Organizational research demonstrates the need to look at both an individual's safety expectations during socialization, and the organizational norms for information seeking. Organizational research also recognizes that not all organizational members have the same access to information (Ford et al., 2014), and that unexpected events create information needs (Miller & Jablin, 1991; Morrison, 2002). Thus, scholars and practitioners ought to consider how structurally imposed information divides, such as a no mobile device policy, influences risk-information seeking frequency and strategy following a workplace injury.

Crisis communication research on information seeking privileges the sources of information people use under uncertain circumstances and how post-crisis resilience is communicatively constructed. Studies on information sources during an emergency prompt practitioners to consider how they can use a combination of sources to meet risk information needs in an organization (e.g., Stephens et al., 2013). Support for demographic predictors of source selection (Sommerfeldt, 2015) also raises questions about the influence of organizational tenure as a predictive characteristic of risk-information seeking. Safety practitioners must engage seasoned employees in ways that capture their attention despite message habituation. Additionally, the recent research on resilience after a crisis (Chewning et al., 2013; Doerfel et al., 2010), demonstrates that individuals return to a stable state, in part, by crafting normalcy through their dialogue with others and through maintaining their communication networks. Scholars looking to understand how workplace injuries disrupt communication and information seeking can use these findings from crisis communication research as entry points for inquiry.

Heath communication research largely focuses on uncertainty as a motivator for information management. Instead of studying information seeking as the inevitable outcome of uncertainty, health scholars recognize that uncertainty may not always yield information seeking (see Brashers, 2001). Extending this assumption to the present area of research provides the space for scholars to explore information management strategies besides active seeking. In particular, Kahlor's (2010) PRISM builds on previous health information management models to identify the communicative drivers of risk-information seeking, and helps frame future empirical work in this area. Furthermore, safety practitioners must evaluate their employees' perceived cost of enacting certain protective health behaviors at work. Given that the provision of protective equipment and requisite safety knowledge does not guarantee safety behaviors (Scott & Trethewey, 2008), it is imperative for practitioners to address workarounds and group norms that increase workplace risk.

In sum, the research from seemingly disparate areas of communication research helps highlight the communicative processes influencing risk-information seeking. Real (2010, p. 426) recognizes that, "research in this arena will require imagination in conjunction with a spirit of collaboration in working with scholars in other disciplines." As such, it is fitting that this chapter extends Real's (2010) call for more studies on occupational risk-information seeking by drawing upon relevant organizational, crisis, and health research on information seeking. From these contexts, we enhance our understanding of the information seeking process under uncertain circumstances. Considering the prevalence of workplace injuries, especially within high-reliability organizations, more knowledge is needed on the communicative processes influencing employee risk-information seeking. Future research on employee risk-information seeking must test and extend the findings from these separate areas of communication research in an effort to assist practitioners in creating workplace environments where barriers to risk information are effectively managed.

Discussion Questions

1. How can organizations encourage employees to seek information about their health and safety at work?
2. When an employee is injured on the job and returns to work, what communicative behaviors would indicate they are resilient?
3. How could prior work experiences impede resilience-building efforts?
4. Are there ways for organizations to gain the attention of employees who have become habituated to certain workplace risks or hazards?
5. What are some possible avenues for interdisciplinary research on the topic of workplace safety?

References

Afifi, W. A., & Afifi, T. D. (2009). Avoidance among adolescents in conversations about their parents' relationship: Applying the theory of motivated information management. *Journal of Social and Personal Relationships, 26,* 488–511. https://doi.org/10.1177/0265407509350869

Afifi, W. A., & Weiner, J. L. (2004). Toward a theory of motivated information management. *Communication Theory, 14*, 167–190. https://doi.org/10.1111/j.1468-2885.2004.tb00310.x

Ajzen, I. (1991). The theory of planned behavior. *Organizational Behavior and Human Decision Processes, 50*, 179–211. https://doi.org/10.1016/0749-5978(91)90020-T

Ashford, S. J., & Cummings, L. L. (1983). Feedback as an individual resource: Personal strategies of creating information. *Organizational Behavior and Human Performance, 32*, 370–398. https://doi.org/10.1016/0030-5073(83)90156-3

Ashford, S. J., & Cummings, L. L. (1985). Proactive feedback seeking: The instrumental use of the information environment. *Journal of Occupational Psychology, 58*, 67–79. https://doi.org/10.1111/j.2044-8325.1985.tb00181.x

Austin, L., Liu, B. F., & Jin, Y. (2012). How audiences seek out crisis information: Exploring the social-mediated crisis communication model. *Journal of Applied Communication Research, 40*, 188–207. https://doi.org/10.1080/00909882.2012.654498

Barbour, J. B., Rintamaki, L. S., Ramsey, J. A., & Brashers, D. E. (2012). Avoiding health information. *Journal of Health Communication, 17*, 212–229. https://doi.org/10.1080/10810730.2011.585691

Barling, J., Loughlin, C., & Kelloway, E. K. (2002). Development and test of a model linking safety-specific transformational leadership and occupational safety. *Journal of Applied Psychology, 87*, 488–496. https://doi.org/10.1037/0021-9010.87.3.488

Bennett, N., Herold, D. M., & Ashford, S. J. (1990). The effects of tolerance for ambiguity on feedback-seeking behaviour. *Journal of Occupational Psychology, 63*, 343–348. https://doi.org/10.1111/j.2044-8325.1990.tb00535.x

Berger, C. R., & Calabrese, R. J. (1975). Some explorations in initial interaction and beyond: Toward a developmental theory of interpersonal communication. *Human Communication Research, 1*, 99–112. https://doi.org/10.1111/j.1468-2958.1975.tb00258.x

Bigman, C. A. (2014). Social comparison framing in health news and its effect on perceptions of group risk. *Health Communication, 29*, 267–280. https://doi.org/10.1080/10410236.2012.745043

Brashers, D. E. (2001). Communication and uncertainty management. *Journal of Communication, 51*, 477–497. https://doi.org/10.1111/j.1460-2466.2001.tb02892.x

Brashers, D. E., Goldsmith, D. J., & Hsieh, E. (2002). Information seeking and avoiding in health contexts. *Human Communication Research, 28*, 258–271. https://doi.org/10.1093/hcr/28.2.258

Brashers, D. E., Neidig, J. L., & Goldsmith, D. J. (2004). Social support and the management of uncertainty for people living with HIV or AIDS. *Health Communication, 16*, 305–331. https://doi.org/10.1207/S15327027HC1603_3

Buzzanell, P. M. (2010). Resilience: Talking, resisting, and imagining new normalcies into being. *Journal of Communication, 60*, 1–14. https://doi.org/10.1111/j.1460-2466.2009.01469.x

Case, D. O., Andrews, J. E., Johnson, J. D., & Allard, S. L. (2005). Avoiding versus seeking: The relationship of information seeking to avoidance, blunting, coping, dissonance, and related concepts. *Journal of the Medical Library Association, 93*, 353–362. www.ncbi.nlm.nih.gov/pmc/articles/PMC1175801/

Casey, M. K., Miller, V. D., & Johnson, J. R. (1997). Survivors' information seeking following a reduction in workforce. *Communication Research, 24*, 755–781. https://doi.org/10.1177/0093650297024006007

CDC (2011). CDC health disparities and inequalities report: United States 2011. *Morbidity and Mortality Weekly Report, 60*, 1–114. Retrieved from: www.cdc.gov/mmwr/pdf/other/su6001.pdf

Chewning, L. V., Lai, C. H., & Doerfel, M. L. (2013). Organizational resilience and using information and communication technologies to rebuild communication structures. *Management Communication Quarterly, 27*, 237–263. https://doi.org/10.1177/0893318912465815

Choi, J., & Chung, W. (2012). Analysis of the interactive relationship between apology and product involvement in crisis communication: An experimental study on the Toyota recall crisis. *Journal of Business and Technical Communication, 27*, 3–31. https://doi.org/10.1177/1050651912458923

Cooper, M. D., & Phillips, R. A. (2004). Exploratory analysis of the safety climate and safety behavior relationship. *Journal of Safety Research, 35*, 497–512. https://doi.org/10.1016/j.jsr.2004.08.004

DeJoy, D. M. (2005). Behavior change versus culture change: Divergent approaches to managing workplace safety. *Safety Science, 43*, 105–129. https://doi.org/10.1016/j.ssci.2005.02.001

Dembe, A. E. (2001). The social consequences of occupational injuries and illnesses. *American Journal of Industrial Medicine, 40*, 403–417. https://doi.org/10.1002/ajim.1113

Doerfel, M. L., Lai, C.-H., & Chewning, L. V. (2010). The evolutionary role of interorganizational communication: Modeling social capital in disaster contexts. *Human Communication Research, 36*, 125–162. https://doi.org/10.1111/j.1468-2958.2010.01371.x

Driscoll, T. R., Harrison, J. E., Bradley, C., & Newson, R. S. (2008). The role of design issues in work-related fatal injury in Australia. *Journal of Safety Research, 39*, 209–214. https://doi.org/10.1016/j.jsr.2008.02.024

Fernández-Muñiz, B., Montes-Peón, J. M., & Vázquez-Ordás, C. J. (2007). Safety culture: Analysis of the causal relationships between its key dimensions. *Journal of Safety Research, 38*, 627–641. https://doi.org/10.1016/j.jsr.2007.09.001

Folkman, S., & Lazarus, R. S. (1980). An analysis of coping in a middle-aged community sample. *Journal of Health and Social Behavior, 21*, 219–239. https://doi.org/10.2307/2136617

Ford, J. L., & Stephens, K. K. (2018). Pairing organizational and individual factors to improve employees' risk responsiveness. *Management Communication Quarterly, 32*, 504–533. https://doi.org/10.1177/0893318918774418

Ford, J. L., Stephens, K. K., & Ford, J. S. (2014). Digital restrictions at work: Exploring how selectively exclusive policies affect crisis communication. *International Journal of Information Systems for Crisis Response and Management, 6*, 19–28. https://doi.org/10.4018/IJISCRAM.2014100102

Freimuth, V. S., Stein, J. A., & Kean, T. J. (1989). *Searching for health information: The Cancer Information Service model*. Philadelphia, PA: University of Pennsylvania Press.

Granovetter, M. S. (1973). The strength of weak ties. *American Journal of Sociology, 78*, 1360–1380. https://doi.org/10.1086/225469

Griffin, R. J., Dunwoody, S., & Neuwirth, K. (1999). Proposed model of the relationship of risk information seeking and processing to the development of preventive behaviors. *Environmental Research, 80*, 230–245. https://doi.org/10.1006/enrs.1998.3940

Griffin, R. J., Neuwirth, K., Dunwoody, S., & Giese, J. (2004). Information sufficiency and risk communication. *Media Psychology, 6*, 23–61. https://doi.org/10.1207/s1532785xmep0601_2

Hines, S. C. (2001). Coping with uncertainties in advance care planning. *Journal of Communication, 51*, 498–513. https://doi.org/10.1111/j.1460-2466.2001.tb02893.x

Hines, S. C., Babrow, A. S., Badzek, L., & Moss, A. (2001). From coping with life to coping with death: Problematic integration for the seriously ill elderly. *Health Communication, 13*, 327–342. https://doi.org/10.1207/S15327027HC1303_6

Jablin, F. M. (2001). Organizational entry, assimilation, and disengagement/exit. In F. M. Jablin & L. L. Putnam (Eds.), *The new handbook of organizational communication: Advances in theory, research, and methods* (pp. 732–818). Thousand Oaks, CA: Sage.

Johnson, J. D. (1997). *Cancer-related information seeking.* Cresskill, NJ: Hampton Press.

Kahlor, L. (2010). PRISM: A planned risk information seeking model. *Health Communication, 25*, 345–356. https://doi.org/10.1080/10410231003775172

Kassing, J. W. (2000). Investigating the relationship between superior-subordinate relationship quality and employee dissent. *Communication Research Reports, 17*, 58–69. https://doi.org/10.1080/08824090009388751

Katz, E., Blumler, J. G., & Gurevitch, M. (1973). Uses and gratifications research. *Public Opinion Quarterly, 37*, 509–523. https://doi.org/10.1086/268109

Kirsh, B., Slack, T., & King, C. A. (2012). The nature and impact of stigma towards injured workers. *Journal of Occupational Rehabilitation, 22*, 143–154. https://doi.org/10.1007/s10926-011-9335-z

Kramer, M. (2010). *Organizational socialization: Joining and leaving organizations* (Vol. 6). Malden, MA: Polity Press.

Kramer, M. W., Dougherty, D. S., & Pierce, T. A. (2004). Managing uncertainty during a corporate acquisition. *Human Communication Research, 30*, 71–101. https://doi.org/10.1111/j.1468-2958.2004.tb00725.x

Kramer, M. W., & Miller, V. D. (2014). Socialization and assimilation: Theories, processes, and outcomes. In L. L. Putnam, & D. K. Mumby (Eds.), *The new handbook of organizational communication: Advances in theory, research, and methods* (pp. 535–547). Thousand Oaks, CA: Sage Publications.

Lambert, S. D., & Loiselle, C. G. (2007). Health information seeking behavior. *Qualitative Health Research, 17*, 1006–1019. https://doi.org/10.1177/1049732307305199

Lewis, L. K. (1999). Disseminating information and soliciting input during planned organizational change: Implementers' targets, sources, and channels for communicating. *Management Communication Quarterly, 13*, 43–75. https://doi.org/10.1177/0893318999131002

Lewis, L. K., Schmisseur, A. M., Stephens, K. K., & Weir, K. E. (2006). Advice on communicating during organizational change: The content of popular press books. *Journal of Business Communication, 43*, 113–137. https://doi.org/10.1177/0021943605285355

Liu, B. F., Austin, L., & Jin, Y. (2011). How publics respond to crisis communication strategies: The interplay of information form and source. *Public Relations Review, 37*, 345–353. https://doi.org/10.1016/j.pubrev.2011.08.004

Mansfield, E., Stergiou-Kita, M., Kirsh, B., & Colantonio, A. (2014). After the storm: The social relations of return to work following electrical injury. *Qualitative Health Research, 24*, 1183–1197. https://doi.org/10.1177/1049732314545887

Mark, G., Al-Ani, B., & Semaan, B. (2009). Resilience through technology adoption: Merging the old and the new in Iraq. In *Proceedings of the 27th International Conference on Human Factors in Computing Systems*, (pp. 689–698). New York: ACM Press

Miller, S. M. (1987). Monitoring and blunting: Validation of a questionnaire to assess styles of information seeking under threat. *Journal of Personality and Social Psychology, 52*, 345–353. https://doi.org/10.1037/0022-3514.52.2.345

Miller, V. D., & Jablin, F. M. (1991). Information seeking during organizational entry: Influences, tactics, and a model of the process. *Academy of Management Review, 16,* 92–120. https://doi.org/10.5465/AMR.1991.4278997

Morrison, E. W. (2002). Information seeking within organizations. *Human Communication Research, 28,* 229–242. https://doi.org/10.1111/j.1468-2958.2002.tb00805.x

Morrison, E. W., & Cummings, L. L. (1992). The impact of feedback diagnosticity and performance expectations on feedback seeking behavior. *Human Performance, 5,* 251–264. https://doi.org/10.1207/s15327043hup0504_1

Myers, S. A., Cranmer, G. A., Goldman, Z. W., Sollitto, M., Gillen, H. G., & Ball, H. (2015). Differences in information seeking among organizational peers: Perceptions of appropriateness, importance, and frequency. *International Journal of Business Communication, 55,* 30–43. https://doi.org/10.1177/2329488415573928

Paul, P. S., & Maiti, J. (2007). The role of behavioral factors on safety management in underground mines. *Safety Science, 45,* 449–471. https://doi.org/10.1016/j.ssci.2006.07.006

Perry, S. D. (2007). Tsunami warning dissemination in Mauritius. *Journal of Applied Communication Research, 35,* 399–417. https://doi.org/10.1080/00909880701611060

Poole, M. S. (2011). Communication. In S. Zedeck (Ed.), *APA handbook of industrial and organizational psychology* (Vol. 3, pp. 249–270). Washington, DC: APA.

Procopio, C. H., & Procopio, S. T. (2007). Do you know what it means to miss New Orleans? Internet communication, geographic community, and social capital in crisis. *Journal of Applied Communication Research, 35,* 67–87. https://doi.org/10.1080/00909880601065722

Ramanadhan, S., & Viswanath, K. (2006). Health and the information nonseeker: A profile. *Health Communication, 20,* 131–139. https://doi.org/10.1207/s15327027hc2002_4

Real, K. (2008). Information seeking and workplace safety: A field application of the risk perception attitude framework. *Journal of Applied Communication Research, 36,* 339–359. https://doi.org/10.1080/00909880802101763

Real, K. (2010). Health-related organizational communication: A general platform for interdisciplinary research. *Management Communication Quarterly, 24,* 457–464. https://doi.org/10.1177/0893318910370270

Scott, C. W., & Trethewey, A. (2008). Organizational discourse and the appraisal of occupational hazards: Interpretive repertoires, heedful interrelating, and identity at work. *Journal of Applied Communication Research, 36,* 298–317. https://doi.org/10.1080/00909880802172137

Seeger, M. W., Sellnow, T. L., & Ulmer, R. R. (2003). *Communication and organizational crisis.* Westport, CT: Praeger.

Sellnow, T. L., Seeger, M. W., & Ulmer, R. R. (2002). Chaos theory, informational needs, and natural disasters. *Journal of Applied Communication Research, 30,* 269–292. https://doi.org/10.1080/00909880216599

Sommerfeldt, E. J. (2015). Disasters and information source repertoires: Information seeking and information sufficiency in postearthquake Haiti. *Journal of Applied Communication Research, 43,* 1–22. https://doi.org/10.1080/00909882.2014.982682

Spence, P. R., Lachlan, K. A., & Burke, J. A. (2011). Differences in crisis knowledge across age, race, and socioeconomic status during Hurricane Ike: A field test and extension of the knowledge gap hypothesis. *Communication Theory, 21,* 261–278. https://doi.org/10.1111/j.1468-2885.2011.01385.x

Spence, P. R., Lachlan, K. A., & Griffin, D. R. (2007). Crisis communication, race, and natural disasters. *Journal of Black Studies, 37*, 539–554. https://doi.org/10.1177/0021934706296192

Spence, P. R., Westerman, D., Skalski, P. D., Seeger, M., Sellnow, T. L., & Ulmer, R. R. (2006). Gender and age effects on information-seeking after 9/11. *Communication Research Reports, 23*, 217–223. https://doi.org/10.1080/08824090600796435

Spence, P. R., Westerman, D., Skalski, P. D., Seeger, M., Ulmer, R. R., Venette, S., & Sellnow, T. L. (2005). Proxemic effects on information seeking after the September 11 attacks. *Communication Research Reports, 22*, 39–46. https://doi.org/10.1080/08824090052000343507

Stephens, K. K., Barrett, A. K., & Mahometa, M. J. (2013). Organizational communication in emergencies: Using multiple channels and sources to combat noise and capture attention. *Human Communication Research, 39*, 230–251.

Stephens, K. K., & Malone, P. (2010). New media for crisis communication: Opportunities for technical translation, dialogue, and stakeholder responses. In W. T. Coombs & S. J. Holladay (Eds.), *The handbook of crisis communication* (pp. 381–395). Malden, MA: Blackwell Publishing Ltd.

Sunnafrank, M. (1986). Predicted outcome value during initial interactions: A reformulation of uncertainty reduction theory. *Human Communication Research, 13*, 3–33. https://doi.org/10.1111/j.1468-2958.1986.tb00092.x

Susskind, A. M. (2007). Downsizing survivors' communication networks and reactions: A longitudinal examination of information flow and turnover intentions. *Communication Research, 34*, 156–184. https://doi.org/10.1177/0093650206298068

Taylor-Clark, K. A., Viswanath, K., & Blendon, R. J. (2010). Communication inequalities during public health disasters: Katrina's wake. *Health Communication, 25*, 221–229. https://doi.org/10.1080/10410231003698895

Trethewy, R., & Atkinson, M. (2003). Enhanced safety, health and environmental outcomes through improved design. *Journal of Engineering, Design and Technology, 1*, 187–201. https://doi.org/10.1108/eb060897

U.S. Bureau of Labor Statistics. (2016). *National census of fatal occupational injuries in 2015* (USDL Publication No. 16–2304). Retrieved from www.bls.gov/news.release/pdf/cfoi.pdf

U.S. Bureau of Labor Statistics. (2017). *Labor force statistics from the current population survey*. Retrieved from www.bls.gov/cps/cpsaat47.htm

Vinodkumar, M. N., & Bhasi, M. (2009). Safety climate factors and its relationship with accidents and personal attributes in the chemical industry. *Safety Science, 47*, 659–667. https://doi.org/10.1016/j.ssci.2008.09.004

Vredenburgh, A. G. (2002). Organizational safety: Which management practices are most effective in reducing employee injury rates? *Journal of Safety Research, 33*, 259–276. https://doi.org/10.1016/S0022-4375(02)00016-6

Watson, B. R. (2015). Is Twitter an alternative medium? Comparing Gulf Coast Twitter and newspaper coverage of the 2010 BP oil spill. *Communication Research, 43*, 647–671. https://doi.org/10.1177/0093650214565896

Weick, K. E., & Sutcliffe, K. M. (2007). *Managing the unexpected: Resilient performance in an age of uncertainty* (Vol. 2). San Francisco, CA: John Wiley & Sons.

Westerman, D., Spence, P. R., & Van Der Heide, B. (2014). Social media as information source: Recency of updates and credibility of information. *Journal of Computer-Mediated Communication, 19*, 171–183. https://doi.org/10.1111/jcc4.12041

Wiehe, S. E., Rosenman, M. B., Wang, J., & Fortenberry, J. D. (2010). Disparities in chlamydia testing among young women with sexually transmitted infection symptoms. *Sexually Transmitted Diseases, 37,* 751–755. https://doi.org/10.1097/OLQ. 0b013e3181e50044

Witte, K. (1992). Putting the fear back in fear appeals: The extended parallel process model. *Communication Monographs, 59,* 329–349. https://doi.org/10.1080/03637759209376276

Zoller, H. M. (2003). Health on the line: Identity and disciplinary control in employee occupational health and safety discourse. *Journal of Applied Communication Research, 31,* 118–139. https://doi.org/10.1080/0090988032000064588

5

TROUBLE AT 30,000 FEET

Twitter Response to United Airlines' PR Crises

Michael A. Cacciatore, Sungsu Kim, and Dasia Danzy

United Airlines in Trouble… Twice

The spring of 2017 was not a great time for United Airlines. Within a three-week period beginning in March the airline found itself embroiled in controversies that drew the ire of many in the public. The first incident occurred in Denver, CO on Sunday, March 26, 2017, when reports surfaced that two teenage girls were being denied access to a flight to Minneapolis, MN due to improper dress (Stack, 2017). The issue was first brought to the public's attention via a tweet sent out by Shannon Watts, a former public relations executive and the founder of the gun safety organization "Moms Demand Action," who happened to be at the airport at the same time. In a series of three tweets Watts wrote:

1. A united gate agent isn't letting girls in leggings get on a flight from Denver to Minneapolis because spandex is not allowed?
2. She's forcing them to change or put dresses on over leggings or they can't board. Since when does @united police women's clothing?
3. Gate agent for flt 215 at 7:55. Said she doesn't make the rules, just follows them. I guess @united not letting women wear athletic wear? (Friedman, 2017)

Lost in these tweets was the fact the girls were not regular customers on the flight. Rather, they were flying as "pass travelers" or "pass riders," a special form of travel that is available to United Airlines employees and their families, and one that allows those passengers to fly for free or severely discounted rates on a stand-by basis (Silva, 2017). For this special form of travel, pass travelers are expected to abide by the United Airlines dress code. "The passengers this morning were United pass riders who were not in compliance with our dress code policy for company benefit travel," the company wrote on Twitter, also adding that "[c]asual attire is allowed as long as it looks neat and is in good taste for the local environment" (Stack, 2017).

The company's passenger contract states that it can deny service to pass travelers who are "barefoot or not properly clothed," but fails to elaborate on that description (Lazo, 2017). However, a United spokesperson said that pass travelers are not allowed to wear Lycra and spandex leggings, tattered or ripped jeans, midriff shirts, flip-flops or any article of clothing that shows their undergarments (Stack, 2017). United's stance on the leggings decision did not quell the controversy, and as the hours and days went by many took to Twitter to share their opinions on the airline and their dress code policies, resulting in the issue becoming a trending topic on the social media platform (Gajanan, 2017; Stack, 2017; Whitcomb, 2017; Zoppo, 2017).

By the end of the day on March 27, United Airlines had posted a statement on United Hub, the company's website for news about the airline. The statement clarified that regular customers were allowed to wear to leggings, and only those deemed "pass riders" were expected to follow the dress code that resulted in the young girls being denied access to their flight (United Airlines, 2017). Nevertheless, questions remained about United Airlines' handling of the issue and whether traces of sexism are present in the dress code and its enforcement. While Twitter conversations about the incident continued, United chose not to weigh in any further.

Just a few weeks later, the airline was back in the news, and again, for all the wrong reasons. On April 9, a 69-year-old passenger, David Dao, was forcibly removed from a flight from Chicago, IL to Louisville, KY. Dao sustained serious injuries, including a concussion and a broken nose, after refusing to give up his seat on what was initially described as an overbooked flight (Bromwich, 2017). It was later revealed that the flight was, in fact, not overbooked. Rather, the incident was a result of United "attempting to make seats available for a flight crew... The airline tried to get volunteers, but when no one stepped forward four people were selected and told to leave" (Bromwich, 2017, para. 15).

The video sparked outrage, and many people took to social media to express their opinions. Once again, United Airlines was in full crisis mode. On Monday, April 10, 2017, CEO Oscar Munoz issued a statement that struck a nerve with many. Although the statement issued an apology to other passengers on board, it did not make reference to Dao's treatment that many had witnessed through video accounts of the incident. It also failed to address the serious injuries Dao received when the airline chose to, in the words of United Airlines, "re-accommodate" the doctor. These injuries included a concussion, broken nose, and a potential need for reconstructive surgery (McCann, 2017). The same day, an internal memo from Munoz to United employees, where the CEO pledged support for United's employees and referred to Dr. Dao as "disruptive and belligerent," was leaked to the press, furthering upsetting many in the public (McCann, 2017).

Anger and threats of boycotts began to circulate online and the incident was being widely discussed via social media. By the end of Tuesday, April 11, United Airlines issued another statement that included another apology and acceptance of blame for the incident (McCann, 2017). Munoz followed up the next day, saying he felt "shame" over the video. Finally, on Thursday, Dr. Dao's lawyer and daughter spoke at a news conference, in which they addressed the cruelty of the situation and his current condition. The airline followed with a final statement, promising to improve how they handle situations as such in the future (McCann, 2017).

While there was undeniable anger directed at United via social media as these events unfolded, we know relatively little about the overall content that was being shared, nor how it fluctuated in the immediate aftermath of the two incidents. This analysis aims to fill this gap in the literature by examining Twitter conversation about United Airlines in the aftermath of these two crisis events. Specifically, we explore a census of all Twitter content by means of an automated social media listening platform. Before outlining this approach further, however, we first situate this study in the literature of crisis communication.

Situational Crisis Communication Theory

Situational Crisis Communication Theory (SCCT) is a useful framework to help understand the interplay between organizations and the public in the context of a crisis (Coombs & Holladay, 2002). SCCT is an audience-oriented theory that provides an explanation about how people perceive crisis cases and how they react to an organization based on the organization's crisis response (Coombs, 2010b). The theory assumes that a crisis is a threat to an organization's reputation and that how an organization takes (or denies) responsibility plays a key role in determining public attitudinal and behavioral outcomes (Coombs, 1995; Dowling, 2002; Fombrun & van Riel, 2004).

Given the focus on responsibility, attribution theory (Heider, 1958) is a useful starting point for understanding SCCT, and indeed, has served as a framework for the formation of the theory. Attribution theory helps explain audience reactions to different types of situations through two key concepts: responsibility and controllability. The idea behind attribution theory is that, particularly in the case of negative incidents or crisis events, individuals are likely to make efforts to discern the cause of those events (Coombs, 2010b). Responsibility is attributed to internal (e.g., mistakes made by the organization or individual themselves) or external (e.g., environmental or situational) factors.

The measurement of attribution has been debated by scholars for a number of years, with proposals made for both three-factor (Russell, 1982) and four-factor solutions (McAuley, Duncan, & Russell, 1992). In his development of SCCT, Coombs (1995) settled on a three-factor solution that linked attribution of a crisis event to issues of stability, external control (controllability), and internal control (the locus of control or personal control). Stability refers to the history of the crisis event—that is, whether the crisis event takes place frequently (stable) versus infrequently (unstable). External control is about whether the cause of the event originated from outside forces or not. The locus or personal control refers to the degree to which the actors themselves can control or mitigate the crisis event. This last dimension is said to reflect "intentionality" (Coombs & Holladay, 1996). Together, these three areas influence an audience's evaluation of organizational responsibility—or blame—for a crisis event (Coombs, 1998).

Organizational responsibility is perceived as greatest when the cause of the crisis is stable, external control is low, and personal control is high (Coombs, 1998; Coombs & Holladay, 1996). Under this scenario the organization has a salient crisis history (high stability), is viewed as being the cause of the crisis (low external control), and is viewed as having a high degree of intentionality related to the crisis (high personal control). Taken altogether, the public will view the organization as highly responsible for the crisis event and as failing to take adequate measures to prevent the crisis from happening. In contrast, organizational responsibility is at its lowest when the cause is unstable (i.e., the occurrence of crisis is an exceptional case in the organization's history), external control is high (i.e., outside forces played a significant role in the crisis occurring), and personal control is low (i.e., there is low intentionality).

In SCCT, attribution of responsibility is treated as a core concept in understanding public perception of crisis events and effective crisis coping and messaging strategies for organizations (Coombs, 2010b). In turn, responsibility influences public perception of organizational reputation (Coombs & Holladay, 2002). SCCT establishes a two-step process for evaluating the level of crisis threat to an organization (Coombs, 2010b). First, organizational stakeholders assess initial crisis responsibility; that is, whether, and to what degree, the organization is to blame for the crisis event. This step involves classifying the crisis type for the organization based on the level of responsibility the organization holds and the subsequent threat to their reputation. Three broad categories exist in this process. An organization is classified as "victim," when they are deemed to have low responsibility for the crisis. As a result, the threat is deemed low. When a crisis is classified as an "accident" the organization is deemed to have some responsibility, and therefore there is some threat associated with the crisis. Finally, a crisis is deemed "intentional" when an organization has strong responsibility for the event. As a result, the threat level is high (Coombs, 2010b). While this classification is open to interpretation, United Airlines is solely responsible for their company policies concerning dress code and access to flights, thereby suggesting that these two crisis events are best categorized as "intentional" in nature.

At the second step, crisis history (whether the organization has had a similar crisis event in the past) and prior reputation (the general state of relationship between an organization and stakeholders) are examined as possible intensifying factors (Coombs, 2010b). When the organization lacks prior crises and has a favorable reputation, a halo effect can occur, which can insulate the organization from major negative impacts on public opinion from a crisis event (Coombs & Holladay, 2006). On the other hand, a so-called "Velcro effect" can occur when a negative reputation or a history of past crises accompany an organization, leading to an intensified public response to the crisis event (Coombs & Holladay, 2001). Particularly with the incident involving Dr. Dao, we can reasonably anticipate a Velcro effect to emerge in online discussions of the airline, whereby attentive publics discuss the "re-accommodation" in light of the leggings incident that occurred almost immediately prior.

Situational Crisis Communication Theory and Messaging

Once the context is set, messaging becomes especially important. According to the theory, instructing and adjusting information need to be the first component of any crisis response (Coombs, 2012). Instructing information tells stakeholders how to physically protect themselves from a crisis event. This might include product recall information or evacuation guidelines in the case of accidents (Coombs, 2010a, 2010b). Adjusting information helps individuals psychologically cope with a crisis. Adjusting information takes the form of expressions of concern or sympathy, or corrective actions to prevent repetitive occurrences of the crisis in the future (Coombs, 2012; Sturges, 1994). Only then, after instructing and adjusting information has been shared with the public, are crisis response strategies used to protect an organization from broader reputational damage the crisis may inflict (Coombs, 2010b).

SCCT proposes that an organization's crisis response strategies need to align with the levels of perceived responsibility and threat faced by the organization (Coombs & Holladay, 2002). The crisis response strategies proposed by SCCT are divided into three primary areas (deny, diminish, and rebuild) and one supplementary area (reinforcing) (Coombs, 2010b). Deny strategies insist that the organization is not responsible for the crisis event by denying responsibility, attacking the accuser, or engaging in scapegoating. Diminish strategies seek to lessen the organization's perceived crisis responsibility and/or the perceived seriousness of the crisis, possibly through justification or excuses. Rebuild strategies are used to enhance the reputation of the organization, and may involve some form of compensation or apology. As a secondary strategy to corroborate the other three primary approaches, reinforcing strategies attempt to use positive information, including reminders of past favorable behaviors, portrayals of victimhood, or ingratiating themselves to the audience. These secondary strategies are used in limited times when the organization has a reservoir of previous good work to share with audiences (Coombs, 2010a, 2010b).

Particular communication approaches work better in different circumstances. For example, in the cases of victim crisis types (e.g., natural disasters, rumors, and workplace violence) with no intensifying factors (i.e., no prior crisis history and the absence of a negative reputation), the mere reliance on instructing and adjusting information might be sufficient as an approach to crisis management, while an accident situation for an organization with a history of crises and/or a negative public reputation is typically better served by issuing an apology and making restitution, in order to reduce reputational damage for the company (Coombs, 2007, 2010b). In the simplest sense, as organizational responsibility increases, the crisis response should be less defensive and focused on organization interests and more accommodative and focused on victim concerns (Coombs & Holladay, 2002).

In the analyses that follow, we examine public response to United's two crisis events. We do so by exploring both the sentiment of tweets, as well as the topic areas touched upon in those tweets. Where appropriate, we also discuss public Twitter response within the context of United's statements on the crisis events.

Method

This study analyzed Twitter conversations about United Airlines using the social media monitoring and analytics platform ForSight by Crimson Hexagon. In taking this approach, we followed the strategy of Su et al. (2017). Specifically, we first utilized Boolean search logic to build a list of keywords designed to capture the universe of United Airlines-related content during the two specific time periods where United had well-known and publicized crisis situations in 2017.[1] This was an iterative process whereby an initial set of keywords were employed and then analyzed in terms of their fit for the topic of interest by examining the tweets that this keyword search captured. If more than one in 20 of the captured tweets were deemed to be unrelated to United Airlines, the keywords were refined in order to better hone in on the topic. The most notable hurdles in this process were the volume of content associated with soccer clubs (e.g., "#UNITED! I am supporting Manchester United when they take on Chelsea – LIVE on SuperSport. #SSFootball #PL https://t.co/uErlL-Hj0F4"), given the large number of teams that employ "United" as a team moniker, and the volume of content that generically mentioned the United States (e.g., "@TwitchOhio Hot News Just See and Enjoy #USA #United States" or "MoneyGram and Ant Financial Enter Into Amended Merger Agreement #Stocks #United States #NASDAQ 100 Components"). Thus, we used the AND NOT function of our Boolean search to reduce the number of tweets referencing these two topic areas. Our final keyword search is as follows: (@ united OR #united OR "united airlines" OR "united air" OR "united airline") AND NOT ("#United States" OR "United States" OR "ATL United" OR "DC United" OR "Minnesota United" OR "Manchester United" OR "West Ham United" OR "Newcastle United" OR "Leeds United" OR "Rother- ham United" OR "Peterborough United" OR "Sheffield United" OR "Scun- thorpe United" OR "Colchester United" OR "Oxford United" OR "Carlisle United" OR "Cambridge United" OR "Hartlepool United" OR "Southend United").

Next, a monitor (the term Crimson Hexagon uses to describe a set of anal- yses) was set up for each of the two crisis events. We focused on the specific crisis events by examining an eight-day period of time beginning the day of the crisis event, and ending one week following. This time period was also settled on via an iterative process. Initial monitors started with a wider period of time, but it was quickly discovered that too large a percentage of off-topic, non-crisis-related content emerged the further we moved from the crisis event. To more directly focus on the immediate aftermath of the crisis, we settled on the eight-day window.

The next step was to train the monitors. The authors trained the monitors themselves during in-person meetings and online meetings, which were facil- itated via screen-sharing technology. The training process involved manually classifying tweets that were randomly selected from publicly available Twitter accounts based on our keyword search into our categories of interest. The au- thors worked collaboratively to develop the coding rules for classifying tweets into the respective categories. This process was informed by an analysis of the risk and crisis literature, but was ultimately an inductive process whereby cate- gories were determined based on an analysis of the content that emerged from our keyword search. We outline these categories shortly.

The creators of ForSight recommend that each category be trained with between 20 and 30 representative and distinct tweets. This practice ensures that the software has a sufficiently reliable number of tweets in each category in order to recognize the underlying linguistic patterns associated with that category and to reliably train the classifier. Once each category contained the recommended number of sample tweets, the three authors independently ex- amined the categories to verify that our initial training process did not result in the same tweet being used multiple times to train a category, and as a final check, we verified that the classified tweets were representative of their respec- tive category. Any disagreements at this stage resulted in further group coding until each and every category contained at least 20 unique tweets that all three group members agreed were emblematic of the category of interest. The av- erage root mean square error of the estimate is approximately 3 percent when there are 100 trained tweets within a monitor (Hopkins & King, 2010). Each of our monitors contained more than 200 tweets when factoring in all categories.

The bulk of the trained categories focused on the type of negative sentiment aimed at United Airlines in the aftermath of the crisis events, although other categories proved necessary. Definitions of each trained category follow, and Table 5.1 contains sample tweets for each category across the two investigated crisis events:

1. The "Negative Financial" category included tweets mentioning the financial loss that had followed or would be expected to follow the crisis event. Pending lawsuits and drops in stock prices were particularly common in this category. This category failed to materialize for the leggings crisis.
2. The "Negative Boycott" category included tweets that specifically mentioned no longer flying United Airlines, or tweets urging others to boycott or otherwise avoid the airline.
3. The "Negative Firing" category included tweets that called for anyone at United Airlines to lose their job. The bulk of these calls for firing were aimed at CEO Oscar Munoz, although the employees involved in the incident and others in the company were sometimes included as well.
4. The "Negative Past Crisis" category included tweets that not only criticized United for the re-accommodation crisis, but also linked it to earlier crises, most notably, the leggings crisis. We did not explore this category for the leggings incident due to a lack of relevant content.
5. The "Negative Other Brands/Persons" category included tweets that were not only critical of United Airlines in the aftermath of the crisis, but did so by referencing other brands or personalities, thereby linking them with negative sentiment on Twitter.
6. The "Other Negative Sentiment" category included tweets that criticized United, without any of the attributes noted in the above five categories.
7. The "Positive/Over It" category included tweets that were either supportive of United Airlines, or thought the controversy was blown out of proportion or should otherwise no longer be discussed.
8. The "Neutral News" category included tweets that shared information about the crisis, but in a solely informational capacity. These largely consisted of the sharing of news headlines, without any other form of editorial context.
9. The "Sentiment Unspecified" category included tweets that required clicking on a link in order to adequately understand the message being sent. The bulk of these tweets were negative in tone toward United Airlines, but because ForSight can analyze only the tweet itself, and not any associated links or images, we treated this content as its own category.
10. The "Unrelated to Crisis" category included tweets that were on-topic in terms of being about United Airlines, but were clearly referencing something outside of the crisis event in question.
11. Finally, as part of the training process, coders are required to train an "Off-Topic" category. Our "Off-Topic" category included any tweets that were captured based on our keyword search, but were not about United Airlines.

TABLE 5.1 Sample tweets for each crisis across relevant categories

	Leggings crisis	*Re-accommodation crisis*
Negative Financial	N/A—too few posts for coding	"Saved $800. Cost us $800 million in stock drop. And we're ok with that. #NewUnitedAirlinesMottos #united"
Negative Boycott	"Boycott United Airlines. Men don't wear leggings; ban is discriminatory against women."	"Boycott united airlines this could have been you your brother father grandpa the passenger was in his sixties"
Negative Firing	"@united - United, you are getting KILLED over these leggings. The girls were 10 and 11 years old. ADMIT you were wrong/ FIRE PR TEAM"	"This is a case of airline encouraged & approved police brutality. Everyone involved should be fired including @ united CEO"
Negative Past Crisis	N/A—too few posts for coding	"If only he'd been wearing leggings he'd not been allowed to board in the first place and it wouldn't have kicked off"
Negative Other Brands/ Persons	"@NadineBabu @united united is the worst airline, worse than Spirit even. Sexualzing and shaming kids is not okay."	"@KenPlume Listen, @united you just give that passenger a pepsi and this all blows over."
Other Negative Sentiment	"@united Stop pretending like your policies are somehow acceptable in 2017. They were shit when they were created and they're shit now."	"That's scary. Being treated like a criminal in front of hundreds of people … when you paid for your seat …"
Positive/Over It	"If you're going to fly for free you should dress to the code of the airlines @united good job standing your ground"	"@united Don't blame the airline because some idiot refused to comply with the flight crew"
Neutral News	"AIRLINE NEWS: Some quick clarity on those United Airlines dress codes – New York Daily News https://t.co/ OZWUhYjcm6 https://t.co/ mt2oGdLc4V"	"Hacker News - United Airlines changes Its Policy on Displacing Customers https://t.co/ TjPSYznbZQ"
Sentiment Unspecified	"RT @rickyftw.@united trigger warning https://t.co/ JTgPPh2GXU"	"We'll give United Airlines credit it looks like they've already found a way to capitalize on all their bad publicity https://t.co/ Om9UcefOCj"

	Leggings crisis	*Re-accommodation crisis*
Unrelated to Crisis	"United Airlines targets Hong Kong-San Francisco route for new aircraft and business class - South China Morning Post https://t.co/yH7gIKh3QQ"	"The Retweet Call Out for @ carterjwm is on!!!!! https://t.co/ CxiKbbAF9C#NuggsForCarter @Wendys @Amazon @Google @Microsoft @United"
Off-Topic	"'But whoever is united with the Lord is one with him in spirit.' -1 Corinthians 6:17 #togetherasone #united #crossingworship'"	"#United bounce back, pounce relegation-bound #Sunderland 3-0 https://t.co/AayNrpCOqk #football"

Data Analysis

For the duration of this paper, we follow the lead of Su et al. (2017) and refer to an opinion as any Twitter discussion that matches the keyword set and date range and fits into one of our trained categories of interest. At this stage it is worth noting that the number of opinions is not an indication of the number of individual tweets. Tweets, even given their character restrictions, can be used to express multiple opinions. Thus, ForSight does not tally up the number of tweets in a given category; rather, it provides estimates of the aggregate proportion of all tweets in each trained category.

Results

United's "Leggings" Crisis

For the purposes of our analysis, we decided to focus on two crisis events that plagued United Airlines during the spring/summer of 2017. Chronologically, the first of these crises was the so-called "leggings" crisis, which we have outlined in the early stages of this chapter.

The leggings crisis produced 233,490 relevant opinions, with approximately 3,000 additional opinions classified as "Off-Topic," meaning they were captured by our keyword search, but did not fit into one of our categories of interest. These numbers provide some assurance that our selection of keywords did a good job of capturing content about this crisis event.

The crisis passed very quickly on Twitter, with a peak of over 130,000 relevant opinions expressed on the day of the crisis event (March 26). That number was cut in half to less than 65,000 relevant opinions by March 27, before falling to approximately 18,000 relevant opinions on March 28, and 6,000 relevant opinions on March 29. Relevant opinions stayed quite consistent over the final four days analyzed, fluctuating between a high of 4,173 relevant opinions on March 31 and a low of 3,147 on April 1. All told, this was not an especially large scandal for United Airlines, as will be made clear when we discuss the results from their second crisis event.

Table 5.2 provides a breakdown of the percentage of captured content that fell into each category of interest. Forty-four percent of expressed relevant opinions fell into the "Other Negative Sentiment" category, meaning they were critical of United, but did not specifically call for a boycott or firings, nor did they reference other notable public figures or brands. The "Neutral News" category was the next most popular category over the duration of the crisis event, with 20 percent of all relevant opinions fitting this description. Posts with unspecified sentiment were next most prolific. Ten percent of all relevant opinions fit this category description, which referred to content where the sentiment could not be determined based on the text alone. These posts oftentimes required a viewer to click on a link, or view an image or video file to understand the tone of the message. While we classify these as having unspecified sentiment, the vast majority of content, when examined in context, was negative toward the airline.

The next largest category was posts unrelated to the crisis. These posts were clearly about United Airlines, but also clearly not about the leggings issue under investigation. About 10% of all relevant opinions fit this description. Interestingly, positive posts were the next biggest category—8 percent of all relevant opinions were positive toward United Airlines. This category was larger than calls for firings (5 percent) and boycotts (2 percent), as well as posts that linked the leggings crisis to other brands or personalities (3 percent).This suggests a significant percentage of the public felt United Airlines was justified in their approach to the leggings situation.

Importantly, there were clear patterns in how public sentiment shifted over the course of the crisis (see Figure 5.1). "Other Negative Sentiment" was at its

TABLE 5.2 Breakdown of expressed opinions by categories of interest

	Leggings crisis	*Re-accommodation crisis*
Negative Financial	N/A—too few posts for coding	7%
Negative Boycott	2%	7%
Negative Firing	5%	2%
Negative Past Crisis	N/A—too few posts for coding	5%
Negative Other Brands/Persons	3%	2%
Other Negative Sentiment	44%	40%
Positive/Over It	8%	3%
Neutral News	20%	7%
Sentiment Unspecified	10%	25%
Unrelated to Crisis	10%	2%

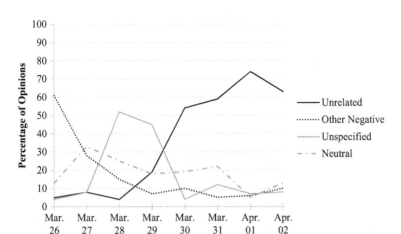

FIGURE 5.1 Proportion of content across four key categories ("Unrelated to Crisis," "Other Negative Sentiment," "Unspecified Sentiment," and "Neutral News") across the eight days examined of the leggings crisis (March 26–April 2, 2017)

peak on the first day of the crisis (March 26), when more than 60 percent of all content fit into that category. This, not surprisingly, suggests that the immediate public reaction on Twitter was a general form of anger directed at United Airlines. Part of this may be attributed to the perceived inadequate response by United, who as noted earlier, failed to issue much of an apology and instead provided excuses for the treatment of the two young girls. If the leggings controversy fits the description of an "intentional" crisis in the public's mind, a hollow apology and justification for the events would be viewed as inadequate by many. By the next day, March 27, the "Other Negative Sentiment" category had decreased to 28 percent of all relevant opinions, and was no longer the most popular form of opinion expression about United Airlines. That distinction fell to the "Neutral News" category, which made up a third of all relevant opinion expression on day two of this crisis event. This is reflective of news organizations having the opportunity to publish accounts of the crisis, and users sharing these accounts via their social networks. By March 28 and 29, posts with unspecified sentiment emerged as most popular, with that category counting 52 percent and 45 percent of all relevant opinions across those two days. A probing of this category suggests that audiences were now processing the crisis in terms of new information that had emerged about the leggings event and were producing often humorous memes or other link- or image-based accounts of the event. Once again, while we classify this content as unclear in terms of sentiment, the accounts were typically negative when examined in the context of attached images and links. Finally, content that was unrelated to the crisis was most popular during the final four days examined here, with that content area containing anywhere from 54 to 74 percent of relevant opinions across each of those four days. This is emblematic of the speed by which the public moved on from this crisis event.

United's "Re-accommodation" Crisis

The "re-accommodation" crisis was the far more prolific of United's two crisis events. Our analysis captured a total of just over 3.7 million opinions that fit our keyword search and date range. Of those, more than 3.5 million opinions were classified as being on-topic, meaning they fit into one of our categories of interest. Fewer than 180,000 of the remaining tweets fit into the "Off-Topic" category, again providing some assurance that our selection of keywords did a good job of capturing content about this crisis event. While that 180,000 number may seem larger, it is worth noting that bots and other users may have noticed the uptick in #united posts and attempted to capitalize on this popularity by means of the hashtag. Such posts were common in our training of the monitor.

The re-accommodation crisis unfolded in a manner similar to the leggings crisis noted above, but with a few exceptions built around the sheer size of the crisis and the timing of the specific crisis event. Most notably, the first day of the crisis was characterized by the smallest volume of tweets (approximately 15,000), although this is easily explained by the fact that the crisis took place in the evening of April 9 leaving little time for public reaction. In contrast, the leggings crisis occurred much earlier in the day, providing more time for publics to react via their social networks. The next day, April 10, the crisis reached its peak with nearly 1.5 million relevant opinions expressed about the issue. As with the leggings crisis, that number steadily declined in the days that followed to just over 1.2 million on April 11, less than 400,000 on April 12, approximately 250,000 on April 13, 120,000 on April 14, 66,000 on April 15, and 54,000 on April 16. While the re-accommodation crisis had greater staying power than the leggings crisis, the general pattern of relevant opinions being cut approximately in half with each passing day was replicated here.

Once again, Table 5.2 provides a breakdown of the percentage of captured content that fell into each category of interest. Approximately 40% of expressed relevant opinions again fell into the "Other Negative Sentiment" category, meaning they were critical of United, but did not reference financial losses for the company, a boycott, possible firings, any past crisis events, such as the leggings incident, or link this crisis to any other notable public figures or brands. The next largest category, with 25 percent of all relevant opinions, was the "Sentiment Unspecified" category. Once again, while we classify these as having unspecified sentiment, the vast majority of content, when examined in context, was negative toward the airline. None of the other categories exceeded

10 percent of all relevant content, with the next highest percentages in the "Negative Financial," "Negative Boycott," and "Neutral News" categories, which each housed 7 percent of all relevant opinions.

Given the timing of Dr. Dao's re-accommodation it is not surprising that opinions unrelated to the crisis dominated the first day's content (see Figure 5.2). Approximately 57 percent of all relevant opinions about United were not re-lated to Dr. Dao being removed from the flight. Most of the remaining 43 percent of opinions fell into the "Other Negative Sentiment" (21 percent) and the "Negative Financial" (20 per cent) categories, suggestive of an initial public condemnation of the incident. By April 10, the day the crisis really started to take off on social media, approximately 60 percent of relevant opinions fit the "Other Negative Sentiment" category, meaning they were generally critical of United. No other category made up even 10 percent of all relevant opinions expressed. In short, there appeared to be public backlash to United's early inad-equate response to the crisis event. By the next day (April 11) there was a split in terms of the most popular content, with 32 percent of all opinions falling into the "Sentiment Unspecified" category and an additional 31 percent falling under the "Other Negative Sentiment" heading. Some of the drop in negative sentiment may be attributed to United finally acknowledging wrongdoing and providing an apology, although it is difficult to say for certain. An examination of posts on this day further revealed that users remained angry at United, but like with the leggings crisis, many were expressing this anger through humor-ous memes and links that came across as innocuous unless a user took the time to specifically click on a link or view an image. Calls for boycotts of United

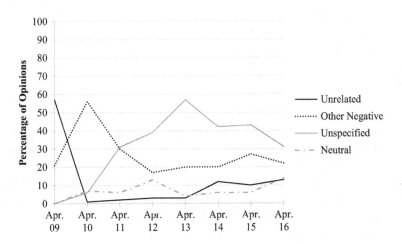

FIGURE 5.2 Proportion of content across four key categories ("Unrelated to Crisis," "Other Negative Sentiment," "Unspecified Sentiment," and "Neutral News") across the eight days examined of the re-accommodation crisis (April 9–16, 2017)

started to make up approximately 10 percent of all relevant posts at this time—a pattern that persisted for several of the remaining days of our analysis. Unspecified sentiment, again, with a heavy focus on humorous memes, remained the most popular form of opinion expression for the duration of our analysis, peaking at a high of 57 percent of all relevant opinions on April 13 and generally containing anywhere from a third to 45 percent of relevant opinions on any given day. "Other Negative Sentiment" was generally the next most popular content category for the remaining days, with anywhere from 20 to 30 percent of all relevant opinions falling under this heading.

Discussion and Practical Implications

In this chapter we explored public sentiment toward United Airlines in the aftermath of two key crisis events during the spring of 2017. Our approach allows for a census of all publicly available tweets that fit our keyword search, thereby providing a complete overview of Twitter discussions around the two crisis events. It also provides valuable information concerning crisis communication and public response to crises in the Web 2.0 environment.

The analysis suggests that there are strong similarities in public response, even with rather dissimilar crisis events. Most notably, general criticisms toward the airline were most common across both monitors, with such sentiment making up approximately 40 percent of all sentiments expressed, regardless of crisis event. Similarly, defenders of the brand fell relatively silent following the two events, with sentiments of support or calls to move on to other issues particularly rare for the re-accommodation crisis. Unspecified sentiment generally followed the same timeline for the two events, as well. Such opinions were generally absent at the very start of the controversy before becoming the most popular form of opinion expression by the third or fourth day of the crisis. Other brands were rather infrequently cited, making up only 3 percent and 2 percent of relevant opinions for the leggings and re-accommodation crises, respectively.

Of course, there were also notable areas of departure between the two crises. Neutral news was more common with the leggings crisis, and made up approximately one of every five opinions expressed on the issue. In contrast, neutral news accounted for less than 10 percent of expressed opinions for the re-accommodation crisis. Calls for firings were more common than calls to boycott United Airlines during the leggings crisis, while the reverse was the case for the re-accommodation crisis. There are a number of potential reasons for this, and certainly the violent nature of the second event makes it a unique crisis situation. This might also suggest that the public processed the leggings controversy as an isolated incident, given that dress codes are not something most flyers encounter or give much thought to when traveling. As such, they might be more supportive of removing the specific employees rather than

boycotting the airline more generally. Conversely, most travelers can probably relate to overbooked flights and attempts to bump passengers to different
flights. Thus, audiences may have viewed this as a broader problem, and one
that goes beyond the removal of a single employee. Indeed, the coding process
suggests that most of the calls for firing following the incident with Dr. Dao
were aimed at the CEO of United Airlines, Oscar Munoz, suggesting that
audiences viewed the incident as a United problem rather than an employee
problem. Unfortunately, a definitive explanation for the above pattern is impossible with the existing data.

This analysis also found at least some evidence for the Velcro effect
(Coombs & Holladay, 2001). Our initial training of the leggings monitor found
virtually no reference to previous crisis events for the airline, and thus, we
excluded the "Negative Past Crisis" category from that analysis. However, it
became quite clear that we needed such a category when building our re-
accommodation monitor. While only 5 percent of all content fit this category,
it does show that a salient crisis event will be referenced in the context of a
wholly new controversy. Of course, the timing of these crises likely played a
role in this process given that the two events occurred within a matter of weeks
of one another. Whether this pattern would have been replicated had months
passed between the two crisis events remains an empirical question worthy of
future research. At the same time, the two crisis events were quite different in
nature, with the first best characterized by issues of gender inequality and dis-
crimination and the second focused on racial inequality and physical violence.
Perhaps a larger Velcro effect would have been present had the two crisis events
focused on more directly comparable issues.

All told, United's response to these two crisis events was not commensu-
rate with the guidelines of SCCT. SCCT sheds lights on the appropriate crisis
communication strategies based on levels of organizational responsibility for a
crisis event. The theory advises against denying responsibility and other forms
of defensive reactions for crises characterized by high levels of organizational
responsibility, which was certainly the case in these two instances. United Airlines' slow movement toward apology and the circulation of the leaked memo
that denied organizational responsibility were incompatible with the key tenets
of the theory and likely played a strong role in the vitriol aimed at the company
on Twitter.

Limitations

It is important to note that this work is not without limitations. To begin, we
have narrowly focused our analysis on Twitter, neglecting other social media
platforms where the crisis events were undoubtedly discussed and debated. This
decision was made for three key reasons. First, with little doubt, Twitter has
emerged as the social media platform of choice for breaking news information

(Shearer & Gottfried, 2017). Thus, it seemed particularly appropriate for understanding public reaction to a pair of evolving news stories. Second, the character restrictions associated with Twitter make it easier for coding content into single categories. In other words, users have less ability and are less inclined to make complex statements on the platform (Crimson Hexagon Help Center, 2017), which enhances our ability to find and classify pure indicators of each of our categories of interest. Finally, how users express themselves is directly tied to the platform they are utilizing. For example, Facebook posts will generally look different than tweets or Instagram captions, given the restrictions associated with different platforms and how users have traditionally communicated content across each (Su et al., 2017). It is therefore recommended that analyses using the ForSight program stick to a single platform at a time when building monitors, rather than simultaneously coding content across multiple platforms (Crimson Hexagon Help Center, 2017; Su et al., 2017). We chose to limit our search to Twitter, rather than bringing additional platforms into the analysis.

Similarly, we decided to focus on an eight-day timeframe for each of the two crisis events, which paints a somewhat narrow picture of public outcry in the aftermath of these two events. Of course, this approach means we have missed several key events that may have reenergized social media discussions of the crises, most notably events like the airline's April 19 decision not to honor a planned promotion for Oscar Munoz (Czarnecki, 2017). Nevertheless, we believe this decision was justified given the speed by which content that was unrelated to the crises emerged within even a few days of their beginnings. While we have certainly missed some public reaction to these two crises, we believe our census of 16 days of content has captured the major narratives that emerged.

Finally, it is worth noting that the ForSight platform only examines the content that is directly contained in social media posts. In other words, the platform does not examine the content contained in links that the post might reference. This played a role in our decision to create a category for content with unclear sentiment, even though the vast majority of content that fit that description ended up being critical of the airline. In this way, our analysis has undoubtedly underestimated the proportion of negative sentiment directed at the airline, and this limitation of the platform should be considered when making sense of our findings.

Discussion Questions

1. In what ways might the results presented here be different had the analysis been focused on Facebook rather than Twitter? What about other social media platforms?
2. Since these crisis events have unfolded, Twitter has expanded the number of allowable characters in tweets to 280. Does that change how audiences are likely to communicate about organizational crisis situations?

3. How should United Airlines have handled these two crisis events? To what degree should expressed anger on Twitter and other social media platforms guide company actions? What other sources of data would you like to see before advising United on how to handle these crisis events?
4. Should United's competitors have weighed in on these controversies? What are the advantages and disadvantages of other companies producing humorous tweets or memes about the troubles United faced?

Note

1 We initially intended on examining a third crisis event for the airline—a May, 2017 incident where a father in a same-sex couple was accused of inappropriately touching his son—however, the event appeared to garner very limited attention on Twitter and was drowned out by other discussions about the airline. Given the lack of content, and our desire to keep a consistent set of keywords across crisis events for comparison purposes, we removed this crisis from the analysis.

References

Bromwich, J. E. (2017). United Airlines passenger was violent in removal, police report says. *New York Times*. Retrieved from: www.nytimes.com/2017/04/25/business/united-airlines-david-dao-passenger.html

Coombs, W. T. (1995). Choosing the right words: The development of guidelines for the selection of the "appropriate" crisis-response strategies. *Management Communication Quarterly, 8*(4), 447–476. https://doi.org/10.1177/0893318995008004003

Coombs, W. T. (1998). An analytic framework for crisis situations: Better responses from a better understanding of the situation. *Journal of Public Relations Research, 10*(3), 177–191. https://doi.org/10.1207/s1532754xjprr1003_02

Coombs, W. T. (2007). Protecting organization reputations during a crisis: The development and application of Situational Crisis Communication Theory. *Corporate Reputation Review, 10*(3), 163–176. https://doi.org/10.1057/palgrave.crr.1550049

Coombs, W. T. (2010a). Crisis communication: A developing field. In R. L. Heath (Ed.), *The Sage handbook of public relations* (pp. 477–488). Thousand Oaks, CA: Sage.

Coombs, W. T. (2010b). Parameters for crisis communication. In W. T. Coombs & S. J. Holladay (Eds.), *The handbook of crisis communication* (pp. 17–53). Malden, MA: Blackwell.

Coombs, W. T. (2012). *Ongoing crisis communication: Planning, managing, and responding.* Thousand Oaks, CA: Sage.

Coombs, W. T., & Holladay, S. J. (1996). Communication and attributions in a crisis: An experimental study in crisis communication. *Journal of Public Relations Research, 8*(4), 279–295. https://doi.org/10.1207/s1532754xjprr0804_04

Coombs, W. T., & Holladay, S. J. (2001). An extended examination of the crisis situations: A fusion of the relational management and symbolic approaches. *Journal of Public Relations Research, 13*(4), 321–340. https://doi.org/10.1207/S1532754XJPRR1304_03

Coombs, W. T., & Holladay, S. J. (2002). Helping crisis managers protect reputational assets: Initial tests of the Situational Crisis Communication Theory. *Management Communication Quarterly, 16*(2), 165–186. https://doi.org/10.1177/089331802237233

Coombs, W. T., & Holladay, S. J. (2006). Unpacking the halo effect: Reputation and crisis management. *Journal of Communication Management, 10*(2), 123–137. https://doi.org/10.1108/13632540610664698

Crimson Hexagon Help Center. (2017). Foundation training series: Opinion monitors. Retrieved from: https://help.crimsonhexagon.com/hc/en-us/articles/115004853103

Czarnecki, S. (2017). Timeline of a crisis: United Airlines. *PR Week*. Retrieved from: www.prweek.com/article/1435619/timeline-crisis-united-airlines

Dowling, G. (2002). *Creating corporate reputations: Identity, image and performance.* New York: Oxford University Press.

Fombrun, C. J., & van Riel, C. B. M. (2004). *Fame and fortune: How successful companies build winning reputations.* New York: Prentice-Hall/Financial Times.

Friedman, M. (2017). The woman who called out United's "sexist" leggings policy speaks out. *Cosmopolitan*. Retrieved from: www.cosmopolitan.com/lifestyle/a9197487/shannon-watts-united-airlines-leggings-op-ed/

Gajanan, M. (2017). United wouldn't let 2 girls on a plane because it apparently has a leggings ban. *Times*. Retrieved from: http://time.com/4713370/united-airlines-leggings-ban/

Heider, F. (1958). *The psychology of interpersonal relations.* New York: Wiley.

Hopkins, D. J., & King, G. (2010). A method of automated nonparametric content analysis for social science. *American Journal of Political Science, 54*(1), 229–247. https://gking.harvard.edu/files/words.pdf

Lazo, L. (2017). Two girls barred from United flight for wearing leggings. *Washington Post*. Retrieved from: www.washingtonpost.com/news/dr-gridlock/wp/2017/03/26/two-girls-barred-from-united-flight-for-wearing-leggings/?utm_term=.206ea817cff0

McAuley, E., Duncan, T. E., & Russell, D. W. (1992). Measuring causal attributions: The revised causal dimension scale (CDSII). *Personality and Social Psychology Bulletin, 18*(5), 566–573. https://doi.org/10.1177/0146167292185006

McCann, E. (2017). United's apologies: A timeline. *New York Times*. Retrieved from: www.nytimes.com/2017/04/14/business/united-airlines-passenger-doctor.html

Russell, D. (1982). The causal dimension scale: A measure of how individuals perceive causes. *Journal of Personality and Social Psychology, 42*, 1137–1145. http://dx.doi.org/10.1037/0022-3514.42.6.1137

Shearer, E., & Gottfried, J. (2017). News use across social media platforms 2017. *Pew Research Center*. Retrieved from: www.journalism.org/2017/09/07/news-use-across-social-media-platforms-2017/

Silva, D. (2017). United Airlines leggings incident shows changing nature of air travel. *NBC News*. Retrieved from: www.nbcnews.com/storyline/airplane-mode/united-airlines-leggings-incident-shows-changing-nature-air-travel-n738926

Stack, L. (2017). After barring girls for leggings, United Airlines defends decision. *New York Times*. Retrieved from: www.nytimes.com/2017/03/26/us/united-airlines-leggings.html

Sturges, D. L. (1994). Communicating through crisis: A strategy for organizational survival. *Management Communication Quarterly, 7*(3), 297–316. https://doi.org/10.1177/0893318994007003004

Su, L. Y.-F., Cacciatore, M. A., Liang, X., Brossard, D., Scheufele, D. A., & Xenos, M. A. (2017). Analyzing public sentiments online: A new content analysis tool combining human- and computer-based coding. *Information, Communication and Society, 20*(3), 406–427. https://doi.org/10.1080/1369118X.2016.1182197

United Airlines. (2017). To our customers … your leggings are welcome! *United Airlines.* Retrieved from: https://hub.united.com/our-customers-leggings-are-welcome-2331263786.html

Whitcomb, D. (2017). How two teens in leggings became a PR mess for United Airlines. *Reuters.* Retrieved from: www.reuters.com/article/us-unitedairlines-leggings/how-two-teens-in-leggings-became-a-pr-mess-for-united-airlines-idUSKBN16Y2HY

Zoppo, A. (2017). Internet erupts after United Airlines boots girls for wearing leggings. *NBC News.* Retrieved from: www.nbcnews.com/news/us-news/internet-erupts-after-united-airlines-boots-girls-wearing-leggings-n738706

6

MOBILE CRISIS COMMUNICATION

Temporality, Rhetoric, and the Case of Wireless Emergency Alerts

Hamilton Bean and Stephanie Madden

In the past decade, high-profile disasters in the United States have sparked innumerable initiatives to improve public alert and warning (Moore, 2011). Many emergency management organizations have turned to short message service (SMS) systems and social media technologies that limit singular alert and warning messages to a few hundred characters or less (Crowe, 2011). Recent advances in mobile communication have provided officials with new ways to alert publics of imminent natural and human-induced disasters, and the use of new media for public warning in the United States is becoming ubiquitous. As part of these trends, the federal wireless emergency alerts (WEA) system, which began its nationwide rollout in 2012 (and is the focus of this chapter), allows mobile device users to receive geo-targeted, text-like messages of imminent threats to their safety via cell broadcast technology. Many people in the United States have received these notifications on their mobile devices. A distinctive alarm sound (similar to a loud buzzer) accompanies the message to capture people's attention. WEA technology ensures that messages can be transmitted during times of network congestion. Authorized alerting authorities (e.g., federal, state, or local emergency managers) can target WEA messages to specific geographic areas (although the precision of geo-targeting varies). Use of the WEA system is voluntary. Mobile device users can opt out of "imminent threat" alerts (opt-in is the default setting for new devices), and commercial wireless carriers participate in WEA via partnership with the Federal Communications Commission (FCC) and the Federal Emergency Management Agency (FEMA).

WEA gained public attention in the wake of both the January 13, 2018 ballistic missile false alarm in Hawaii and the October 2017 Northern California wildfires. The false WEA message issued in Hawaii terrified many residents and tourists. State officials subsequently revised the processes for sending alerts to ensure that at least two people reviewed the message and approved its release. The Northern California wildfires killed 44 people and were the costliest wildfires ever recorded. During one blaze, Sonoma county officials declined to use WEA and instead issued alerts via opt-in systems and landline connections. These alerts and warnings reached only a fraction of the county's residents (St. John, 2017). Sonoma county officials justified their decision to avoid WEA based on the imprecision of its geo-targeting capabilities, "Providing mass information to people not affected could have caused mass traffic backups, which could have impacted emergency service providers and delayed emergency response" (Palomino & Veklerov, 2017, para. 7). California lawmakers subsequently introduced plans to mandate the use of WEA during similar emergencies in the future, as well as establish consistent standards across all 58 counties. These events also spurred emergency managers nationwide to send a letter to the FCC on January 5, 2018 requesting upgrades to the WEA system, including improved geo-targeting, the use of multimedia, many-to-one interactivity (i.e., polling questions that could allow emergency managers to more effectively deploy resources), and multilingual messages (Robert J. Downen, personal communication, January 8, 2018). WEA researchers (including the authors of this chapter) have urged officials to enact similar changes, but wireless carriers have lobbied against upgrades and resisted their implementation due to fears of increased cost, technological challenges, and possible network congestion (Romm, 2017).

Organizations that are authorized to use WEA are thus currently considering whether and how small maps might be embedded within or accompany WEA messages in the future. For example, a map could indicate preferred evacuation routes in order to reduce traffic backups. It is possible, however, that map inclusion could also produce confusion and unwanted outcomes. Recent research that has investigated the possibility of map inclusion within WEA messages has generated mixed findings (Bean et al., 2014, 2015, 2016; Cao, Boruff, & McNeill, 2016, 2017; Casteel & Downing, 2016; Liu et al., 2017; Wood, Bean, Liu, & Boyd, 2015; Wood et al., 2017). In some cases, map inclusion appears to improve message recipient interpretation and decision-making outcomes, but in other cases, map inclusion appears to have little influence (or even negative influence) on people's understanding and decision-making.

These divergent findings invite researchers and practitioners to reconsider the dominant theoretical orientation and methodological tools used to investigate WEA map inclusion specifically and mobile crisis communication generally. The events in Hawaii and California (as well as Texas and elsewhere) indicate that there is an urgent need for innovation in mobile crisis communication research. The aim of this chapter is to explore a rhetorical approach to mobile crisis communication, one that emphasizes context, complexity, and indeterminacy in order to improve public understanding of the benefits and limitations of new media, especially in regards to the timeliness and specificity of messages.

The chapter unfolds in four sections. First, it briefly provides background about the WEA system and discusses mobile crisis communication's dominant socio-psychological orientation and research methods. This section identifies how temporality complicates officials' ability to adequately understand and control public behavior during rapid-onset emergencies. Second, highlighting recent research that investigates possible map inclusion in WEA messages, the chapter explores shifting away from the dominant orientation to consider what a rhetorical approach might offer scholars of new media and mobile crisis communication. Finally, the chapter identifies some practical implications of a rhetorical approach for WEA officials and crisis communicators.

Temporality and Crisis Communication Research Theories and Methods

The WEA system, launched in 2012, limits text-type messages to a mere 90 characters. However, in 2016, the FCC approved the expansion of WEA messages to 360 characters, and gave wireless carriers until 2019 to implement the expansion. WEA can be categorized as an example of new media, in that WEA message reception requires the possession of a WEA-compatible smartphone. The WEA system relies on digital technologies to produce messages (via "authorized alert originators"), distribute messages (via wireless networks), and render messages (via WEA-compatible digital devices). However, WEA messages currently are not interactive and instead mimic broadcast technologies—interactivity being a key characteristic attributed to most new media technologies. Nevertheless, WEA messages are *potentially* interactive in that they can be configured to provide recipients with URLs, polls, or telephone numbers to call for additional and confirming information.

Along with its uncertain status as "new media," WEA's classification as risk and/or crisis communication remains murky. While risk and crisis communication theories have long aided in disaster prevention, mitigation, preparation, response, and recovery, *mobile* crisis communication messages, such as WEA messages, have only recently gained researchers' attention (Bean et al., 2015). This chapter clarifies that WEA messages occupy a unique position at the junction of risk, crisis, and disaster communication. Sellnow, Ulmer, Seeger, and Littlefield (2009) explained that risk communication involves projections about harms that may occur at some future date. Risk communication focuses on how to reduce the likely consequences of that harm based on what is currently known about it. Risk communication also entails long-term planning, message preparation, and the participation of technical experts and scientists. By contrast, crisis communication involves a specific incident that has already occurred and produced harm. Crisis communication often addresses the magnitude, immediacy, duration, cause, and consequences of an incident. It is based on known and unknown dangers, is reactive, and tends to involve authority figures, emergency managers, and technical experts. Coombs (2010, p. 59) acknowledged that disaster communication and crisis communication were "allied fields" but not totally "isomorphic." Coombs instead defined disaster communication as focusing primarily on relief and restoration efforts and post-event coordination among multiple agencies and private-sector organizations. WEA messages reside in a theoretical grey area in Coombs' typology because they address imminent threats to life or property for populations within a specific geographic area for an incident that is just about to occur or is in the process of occurring.

Sellnow and Seeger (2013) have shed light on this theoretical grey area, describing public warning (of which WEA messages are a primary example) as a specialized form of risk communication that contains elements of crisis communication. Public warning models address how crises are detected and how communities are alerted, protected, or evacuated. Here, the focus is on how to adequately notify publics to probable risks and effectively motivate them to take appropriate protective action. Sellnow and Seeger identified Mileti and Sorensen's (1990) "hear-confirm-understand-decide-respond" model as illustrative of public warning theory. In brief, this model accounts for the complex, individual social-psychological responses from the moment of seeing/hearing a warning message through protective action taking. Upon seeing/hearing a message, a recipient will usually search for (i.e., "mill") for additional and confirming information before making an assessment of personalized risk. Once one's level of personalized risk is established, that person will decide whether and how to take protective action, as well as when to initiate that action. Sellnow and Seeger (2013) argued that the Mileti and Sorensen model usefully drew attention to issues of message consistency, repetition, understandability, and credibility. However, Sellnow and Seeger critiqued the model for its imprecision concerning the effectiveness of specific warning messages in generating desired behavioral outcomes.

As co-investigators with Mileti on a U.S. Department of Homeland Security-funded research project that investigated public responses to mock and actual WEA messages, the authors of this chapter, along with colleagues, used the Mileti and Sorensen model to guide an investigation of how audiences made sense of an array of WEA message variables. Qualitative research efforts concentrated on the collection and analysis of think-out-loud and focus group interview data, and the hear-confirm-understand-decide-respond model was used in the development of interview questions that participants could respond to in ways that revealed more in-depth and nuanced information than could be captured solely via survey responses and quantitative analysis (Bean et al., 2014). Nevertheless, the research team was constrained in its efforts by a phenomenon that influences qualitative and quantitative risk, crisis, and disaster research alike: temporality.

Temporality

The experience of time is fundamental to human existence, and issues related to time (i.e., its nature, perception, and experience) are captured under the moniker of "temporality." The moment of a person's decision-making during a disaster situation is fleeting, and researchers cannot observe the cognitive processes involved in message receipt, interpretation, and response. Researchers cannot know for certain why, for example, someone who is instructed to evacuate ultimately decides to do so while another person does not—with the latter placing themselves (and possibly their loved ones) in peril. The relationships among decision-making, observation, and time have limited the ability of crisis communication researchers to uniformly model and accurately and consistently predict public response to emergencies. When researcher observation of public response to a disaster has occurred, what is typically seen is not instantaneous decision-making and action-taking, but a *deferral* of these moments—a process that disaster sociologists have termed *milling*; that is, searching for additional and confirming information (Wood et al., 2012, 2017). This additional and confirming information may be derived from the physical environment, by talking to or observing other people, or via various forms of media. For some people, the process of milling occurs quickly, but for others, milling may be slower—even outlasting the danger itself.

The trajectory of new media development has favored technologies that have maximized the erosion of spatial and temporal constraints, thereby potentially reducing the time needed to mill for additional and confirming information during an emergency. Today, roughly 80 percent of Americans possess, in their pockets, mobile devices that can alert them to impending disasters at speeds that would have been difficult to imagine even a few decades ago. For example, today's mobile-device-enabled earthquake early warning technologies can instruct people to take cover seconds *before* a tremor is felt. Similarly, WEA messages can notify recipients of impending disasters at speeds that render the human propensity to mill a dangerous liability. For example, in 2013, one WEA recipient immediately took action to protect children in her care during a tornado, sheltering them under tables in a cafeteria. Had she milled for additional information, it is likely that some of the children would have been injured from falling debris in the structure they were inside, which was destroyed by the tornado only seconds after the children had crawled under the tables (Saulmon, 2013). The 2017 northern California wildfires also illustrated the deadly consequences of milling. Some of those who hesitated to immediately evacuate were killed by the fast-moving flames (Fuller & Pérez-Peña, 2017). In their study of campus active shooter events, Stephens, Ford, Barrett, and Mahometa (2014) found that text messaging was most frequently used during the attention-getting phase of the emergency. To our knowledge, researchers have not reported *directly* observing response behavior of WEA or text message recipients (however, as discussed below, new media is enabling this type of observation to occur). Researchers have instead relied upon another source of data hampered by temporality: participant recall.

Researchers use field surveys and interviews to reduce the delay between people's actual experience of a disaster and their recollection of their thoughts and actions at the time. Researchers can ask people what they thought about and did before or during a disaster, but the fidelity of people's accounts is inevitably shaped by social desirability and one's awareness and memory of their own thoughts and behaviors. Since researchers may be skeptical of the accuracy of participant recall, they may seek to "triangulate" such material with data derived from public records, field reports, and other sources. Given researchers' inability to recreate experimental or interview conditions wherein participants confront the same life or death circumstances of an actual emergency, data from mock disaster scenarios can only assess participants' behavioral intentions. These temporal constraints are evident in recent studies of map inclusion in mobile alert and warning messages (Bean et al., 2014, 2015, 2016; Cao et al., 2016, 2017; Casteel & Downing, 2016; Liu et al., 2017; Wood et al., 2015, 2017).

Wireless Emergency Alerts and Map Inclusion

One possibility for increasing WEA interactivity is to include maps. Maps can help message recipients understand who should and who should not take protective action in response to a hazard. Of course, not all maps are equally useful, and it is unclear, exactly, how maps might be rendered to best accompany WEA messages. In investigating map elements in disaster communication more broadly, Dransch, Rotzoll, and Poser (2010) used the example of flood risk maps in Saxony, Germany, where online maps are used as a tool for risk communication. The authors argued, "Maps which should improve risk perception have to depict information not only in a vivid manner but also in a suitable complexity. Cartographic rules about comprehensible symbolization, suitable presentation method, and adequate number of presented topics should be applied" (Dransch et al., 2010, p. 300). Also in the context of flood maps, Hagemeier-Klose and Wagner (2009, p. 565) claimed, "A good initial message could be a well-designed flood map which can lead to a high attention level and to further information seeking by the users."

But what defines a well-designed map? To investigate map comprehension, Haynes, Barclay, and Pidgeon (2007) interviewed residents of Monserrat, a Caribbean island where 19 people died from pyroclastic flows that hit evacuation zones when the island experienced heightened volcanic activity in 1997. Using one regular contour map, six 3D maps, and two perspective photographs of Monserrat, the researchers presented a questionnaire to residents that tested their map-reading skills, comprehension, and understanding. The researchers found that most respondents were able to locate the exclusion line, but only two respondents were able to relate the position of this line to the actual terrain. This suggests that using maps without any explanatory information could be problematic due to variation in map-reading literacy.

Researchers have recently begun to explore how a map or other location-related information can best be communicated *along with* crisis messages delivered over mobile devices (Bean et al., 2014, 2015, 2016; Cao et al., 2016, 2017; Casteel & Downing, 2016; Liu et al., 2017; Wood et al., 2015, 2017). This scholarship also has produced divergent findings. Specifically, in their comprehensive testing of WEA messages for a mock nuclear device detonation scenario, Bean et al. (2014) employed experiments, focus groups, and a field survey to investigate multiple aspects of WEA interpretation and response. Regarding map inclusion, the researchers argued:

> there certainly would be a benefit from adding a high-information map to a WEA message. Doing so could help the public interpret and personalize the worded message, which could, in turn, move more people at risk to take protective action.
>
> *(Bean et al., 2014, p. 4)*

Yet, when messages were limited to 280 characters and modified for different hazard scenarios, researchers subsequently found, "None of the map elements tested had a statistically significant effect on message outcomes, and research participants varied widely in their reactions to the tested maps" (Wood et al., 2015, p. i). These researchers concluded, "Maps can be useful in message personalization, but the role they play varies based on message length. [...] Adding maps, as tested here for messages that are 280 characters in length, was not promising, but merits additional research" (Wood et al., 2015, pp. i–ii).

Bean et al.'s (2016) subsequent study of WEA message interpretation relied exclusively on interview and focus group data gathered for these earlier studies. Participants identified reasons for including maps with WEA messages (e.g., "I wish there [...] was a map, which gave you the area [affected]. This one doesn't [have a map], so I'd be kind of confused getting this" (Bean et al., 2016, p. 142). Participants also offered reasons for excluding maps (e.g., "I'm kind of confused on the radiological hazard warning, and then also with the way they highlight the area [on the map]"; Bean et al., 2016, p. 142). Using the same interview and focus group data, combined with experiment and field survey data, Wood et al. (2017) noted that maps were opportunistically used by some participants to justify why they disregarded recommended protective action guidance. Specifically, some participants claimed that they were close enough to the edge of the hazard area to ignore the warning message. Liu et al. (2017) drew upon the same underlying data as the Wood et al. (2017) study, but Liu et al. specifically focused on the issue of map inclusion. Liu et al. (2017) found that the maps used in the study only marginally improved message understanding, but the researchers nevertheless concluded that map comprehension, personalization, and warning compliance were closely related.

Cao and colleagues (2016) focused on the case of wildfire evacuation. Residents in a wildfire prone area of western Australia were given spatial information in maps and text messages and asked to evaluate both for comprehension and risk perception. The researchers found that combining maps with textual information yielded the best outcomes, but not dramatically. Casteel and Downing (2016) used a similar approach in testing map inclusion for weather-related emergency messages, finding that recipient decision-making times were the same across conditions and that none of the other message content measures influenced behavioral outcomes. Thus, overall, the research record indicates that map inclusion may, at various times and under various conditions, (a) improve risk perception and personalization, (b) contribute little to actual behavioral outcomes, or (c) create possibilities for misinterpretation and misunderstanding.

Explaining the Divergent Findings

The temporal challenges associated with direct observation, participant recall, and intention could potentially account for some of the inconsistent and ambiguous findings of the map-related mobile crisis communication research discussed above. Study design and methodological choices could also potentially account for some of the discrepancies. These studies employed a variety of quantitative and qualitative approaches. The type, level, and precision of data collection, analysis, and interpretation necessarily vary. For example, Bean et al.'s (2016) qualitative methods were designed to *emphasize* ambiguity in ways that quantitative methods downplay. Quantitative methods emphasize estimates of reliability, validity, and generalizability that qualitative research generally eschews. These differences, along with variation in the types of hazard scenarios investigated, the types of maps used, the variety of textual information included or not included with the maps, as well as the heterogeneity of the research participants, certainly could account for the divergent findings.

Nevertheless, we use these findings as an invitation to explore a theoretical and methodological approach that has not heretofore been employed in the mobile crisis communication research arena: rhetoric. A rhetorical approach may possess untapped potential for understanding processes of alert and warning message interpretation of response. The chapter next explores how a rhetorical approach can illuminate issues not easily seen using the socio-psychological orientation's focus on cognitive and behavioral variables and control. Socio-psychological and rhetorical approaches may be combined in useful ways in new media and crisis communication research.

From Control to Convergence: Exploring a Rhetorical Approach

At their core, WEA messages specifically, and mobile crisis communication messages generally, are *arguments* sent to audiences to persuade them that an event of importance has occurred and that action is necessary (Sellnow et al., 2009). This message-centered, rhetorical approach views communication as a series of interacting arguments that either encourage or discourage protective-action or risk-reduction behavior. There is a growing recognition that a message-centered approach is important in the broader context of risk communication (Anthony, Sellnow, & Millner, 2013). Thus, this section first explains what a rhetorical approach entails and then explores how it might help innovate mobile crisis communication research.

Sellnow and colleagues (2009) leveraged the classic argumentation perspective of "convergence" in Perelman and Olbrechts-Tyteca's (1969) *The New Rhetoric: A Treatise on Argumentation*, adapting it for a message-centered, risk communication framework. The impetus for *The New Rhetoric* was to provide applied understandings of how people reason to "actions or behaviors" (McKerrow, 1990, p. 19). According to Perelman and Olbrechts-Tyteca (1969, p. 471), convergence occurs when "several distinct arguments lead to a single conclusion." Perelman and Olbrechts-Tyteca observed that arguments interact with each other on multiple levels: (a) between arguments from different sources; (b) between arguments and the context; (c) between arguments and their conclusions; and (d) between arguments in the discourse about the discourse. When arguments about risk converge in ways that conflict with previously held assumptions, audiences will reconsider or adapt their "reality" to account for this change (Venette, 2003). While audiences may recognize the strength of converging arguments, it is also possible for audiences to mistrust messages if they are thought to be offered in ways that lead to a pre-established conclusion (Perelman & Olbrechts-Tyteca, 1969). Therefore, genuine (rather than coerced) convergence relies on spontaneous and emergent elements of arguments that audiences help co-create.

Sellnow and colleagues (2009) theorized risk communication as a series of interacting arguments because assessment of risk generally involves multiple (and often conflicting) messages on any given issue. Considering convergence as the primary objective in risk communication foregrounds the complexity, variability, and indeterminacy of the rhetorical terrain. Because no single argument is likely to immediately win total acceptance, convergence can account for the non-linear ways that audiences make sense of their circumstances during a crisis.

An interacting-arguments perspective thus contributes to our understanding of milling in two novel ways: (a) it views decision-making and action-taking as *rhetorical* processes wherein argument convergence is required for resolution; and (b) it suggests that these rhetorical processes operate *systematically* as audiences rapidly collect and contemplate information from various sources. Instead of viewing milling as a problem of human nature that negates the advantages of rapid warning systems, researchers need to understand the ways that technology can be used in conjunction with this natural response. Rhetoric can help to illuminate the *discursive* dynamics of milling, adding to existing research that focuses mostly on socio-psychological processes.

To test the usefulness of a rhetorical approach in an applied context, Anthony and Sellnow (2011) interviewed Hurricane Katrina survivors to understand what information-seeking and uncertainty-reduction strategies Gulf Coast residents used to make decisions about their disaster response. Focusing specifically on the role of media sources in the convergence process, Anthony and Sellnow found that local media had the greatest influence on residents' perception of the crisis and perceived self-efficacy because it presented practical information about what actions to take (e.g., evacuate or shelter in place). By contrast, national media mostly sensationalized the story and propagated disaster myths, which was perceived as inaccurate and unhelpful by those in the areas affected by the crisis.

Anthony and colleagues (2013) subsequently used the convergence perspective to understand how individuals seek and make sense of information in the midst of an unfolding crisis situation. The authors conducted a series of focus groups comprised of people who observed simulated television reports about a mock food contamination crisis. The results suggested that the argumentation approach provided a useful framework for studying crisis communication, even as audiences received different messages from diverse sources in a condensed timeframe. Importantly, participants discussed avoiding a rush to judgment when considering multiple arguments related to the crisis. In particular, participants discussed remaining "in waiting" (Anthony et al, 2013, p. 360) for the most accurate information about the crisis and the best action to take for protecting themselves. This finding underscores the problem of achieving convergence during a rapid-onset emergency where information is limited, unclear, or inaccurate.

While convergence is a useful concept, it is not easily adapted to the complexities of mobile crisis communication in rapid-onset emergencies. The concept may help explain what occurs rhetorically in the process of milling, but how the concept might be leveraged to produce more effective public alert and warning messages is an open question. The temporal dimension of convergence is ripe for theoretical extension and practical exploration. As Nowotny (1994, p. 7) explained, "Coping with the problem of time often opens up those boundaries of theory which stimulate further development." The next section thus elaborates the social construction of time in order to develop a set of practical implications related to a rhetorical approach.

The Rhetorical Construction of Temporality

Two perspectives currently dominate time research: (a) clock time (also called experienced time) and (b) socially constructed time (also called interpreted time) (Duncheon & Tierney, 2013). Clock time refers to a linear, objective characterization of time (Zerubavel, 1982). This includes a scheduled format where the sequencing of events is vital (Charmaz, 2003). Socially constructed time, however, recognizes that the interpretation of time may not be uniform (Adams, 1988; Hassard, 1990). Although viewed as distinct conceptualizations, there is a relationship between clock time and socially constructed time that is embedded in social situations and structures (Liao et al., 2013). In particular, "all members of a society share a common temporal consciousness; time is a social category of thought, a product of society" (Lee & Liebenau, 2000, p. 46). New media technologies have sparked a third theoretical perspective— "virtual time"—to adapt to the immense compression of time and space in an increasingly networked society (Castells, 2000; Lee & Liebenau, 2000; Wajcman, 2008).

According to Hariman (1989, p. 215), the primary concern of rhetoric is "essentially temporal," and a rhetorical approach views clock time, socially constructed time, and virtual time in relation to the Greek concepts of *chronos* and *kairos*. *Chronos* refers to chronological time that is measurable by clocks, while *kairos* focuses on time episodes of beginning, middle, and end that involve a "living time of intentions and goals" (Jacques, 1982, p. 14). *Kairos* corresponds roughly to the English word "timely." According to Smith (1969), there are three concepts related to *kairos*: (a) timing (the right or opportune time); (b) tension that calls for a decision; and (c) the opportunity to accomplish a purpose. The concept of timing is vital to understanding temporality in a socially constructed or virtual sense (Flaherty, 2011).

Alerts and warnings mostly involve a *kairotic* understanding of time because of the importance of issuing messages at the "right" time. There is a fundamental tension that exists between the specificity and urgency of message delivery. Scott (2006, p. 119) argued that *kairos* always includes an inherent risk, explaining that "kairotic action can be based on the assessment of and attempt to opportunistically control or at least avoid or defend against risk." Viewing time in these terms emphasizes not only alert and warning message issuance but also message reception, as "agency is equally operative regardless of whether someone decides to do something at a particular time or decides not to do something at a particular time" (Flaherty, 2011, p. 88).

Within the context of mobile alerts and warnings, considering timing alone is not enough: organizations issuing these messages must also consider place. There are two distinct Greek notions of place: *chora* and *topos*. *Chora* represents abstract space whereas *topos* refers to concrete and meaningful place (Rämö, 1999). Considering *kairos* along with *topos* results in the concept of *kairotopos*.

Kairotopos is defined as the unification of time and place condensed into a meaningful and concrete whole. Rämö (1999, p. 322) argued that "acting in a kairotopos-sense requires a feature of voluntary action beyond official responsibility that encompasses circumstances that the individual is aware of and a choice is made." Mobile alert and warning can be understood through the lens of *kairotopos* as the intersection between meaningful time and concrete place. Time and place work together, and this emphasis leads us back to the need for innovating mobile crisis communication research in order to support practice. The final section explores the practical implications of a rhetorical perspective that emphasizes *kairotopos*.

Practical Implications

The concept of *kairotopos* offers possibilities for the innovative development of mobile crisis communication research and practice. Mobile devices provide a potential goldmine of data regarding behavioral response to specific WEA messages at specific times and in specific locations. For example, the false alarm WEA issued in Hawaii enables researchers to directly observe the reactions and responses of many people who received the message, as they have uploaded to the Internet videos of themselves receiving the message or commenting on their response only moments after receiving it. Through analysis of these videos, it is possible to determine when a WEA message was broadcast to a given geographic area (or "polygon" in the parlance of WEA notification), pinpoint the geographical locations of the mobile devices and people who received that WEA message, and observe the subsequent movement (or not) of those devices and people.

The Hawaii WEA message included instructions to shelter-in-place; researchers can identify whether people generally stayed put or changed location. Likewise, through device tracking technology, in the future, if an alerting authority issued a WEA message that instructed recipients to evacuate, researchers could analyze which mobile devices moved where, as well as the time it took for those mobile devices to move away from the hazard. Video upload and device-tracking data could then be used by researchers to later ask specific WEA message recipients to account for their milling and decision-making in order to identify the specific *arguments* that influenced message interpretation and response. In this way, mobile crisis communication researchers could use the concept of *kairotopos* to identify "argument clusters" in order to speed convergence. The idea here is to use *kairotopos* to guide the development of (preplanned) arguments addressed to specific audiences. For example, if officials are aware that a group of residents in a certain location are likely to be deeply concerned about the welfare of their livestock during an emergency, officials could plan to issue messages that included compelling arguments addressing livestock protection. Employing the concept of "argument clusters" could subtly change the focus of crisis communication planning and response to better account for rhetorical dynamics.

Tourist populations are another group in which a *kairotopic* approach to mobile crisis communication can help develop more compelling argument clusters. The WEA message polygon can target areas where tourists typically congregate (based on a concentration of hotels and attractions) and refine messages to those populations. For example, rather than simply saying "check local media," more specific information can be given in tourist areas about what the local media are and places to access additional information for those who are unfamiliar with the area.

Rhetoric is ultimately about the use of persuasion, and mobile crisis communication requires persuasive messaging to "form attitudes or induce actions" in recipients (Burke, 1945, p. 41). Of course, even with insights into *kairotopos*, cognitive processes are still unobservable and could only be accounted for in hindsight; yet, in theory, it would be possible for alerting authorities to analyze video and device tracking data to better identify which recipients complied with WEA message instructions and which did not (i.e., those near the edge of the hazard area, those from certain geographic locations, etc.). The technological aspects of *kairotopos*-related research are straightforward. Mobile device tracking is an established industry in the United States (e.g., LocationSmart), and researchers could recruit in advance study participants amenable to client software installation and GPS-enabled device tracking in the case of a rapid onset emergency. The tracking of device data would provide opportunities to observe the actual result of messages, rather than relying on self-reported data. Video analysis and mobile device tracking may be as close as researchers ever get to the witnessing of the decision-making processes of individual WEA message recipients.

Even though such investigations are technologically possible, as currently configured, WEA is not designed to and does not track the location of anyone receiving a WEA message. More to the point, such research methods may risk running afoul of privacy expectations, privacy regulations, the use of mobile data, and fears of government intrusion. There would be difficulty in predicting and controlling which groups of people are likely to experience a rapid-onset emergency (let alone one that included the issuing of a WEA message). Therefore, location tracking remains only on the horizon of possibility for mobile crisis communication research (Bean et al., 2015).

Nevertheless, at its core, a rhetorical approach to mobile crisis communication allows for conceptualizations of risk as socially constructed and as "messy rhetorical enterprises rather than more straightforward scientific ones" (Scott, 2006, p. 119). In other words, a rhetorical approach allows officials and researchers to grasp nuances and complexities of mobile alerts and warnings in ways that other approaches may not. A rhetorical approach also opens up new interdisciplinary research opportunities. A rhetorical approach authorizes data collection elements from qualitative research (e.g., ethnography, interviews, focus groups) and rhetorical criticism (which focuses on arguments and evidence) (Middleton, Senda-Cook, & Endres, 2011). Although traditional approaches to rhetorical criticism typically do not use researcher-facilitated texts as the basis for analysis, Middleton et al. (2011) argued that engaging in social relationships allows for nuanced data regarding the diverse identities and interpretations shaping rhetorical exchanges. For example, given the premise that time is a social construction, rhetorical criticism offers an approach for understanding how interpretations by public safety officials (such as those tasked with sending WEA messages) perform a rhetorical enactment of time as part of a persuasive appeal (Madden, 2017).

Finally, a rhetorical approach to mobile crisis communication allows researchers to dig deeper into the underlying processes and assumptions that are often taken for granted in dominant socio-psychological approaches. Future research can gather information from the field about the decision-making processes involved when issuing and receiving mobile alerts and warnings, showing how officials and publics rhetorically construct their interpretation of time and place and the constraints they face in their day-to-day "living rhetorics" (Middleton et al., 2011). This reframing of the text–audience relationship allows for an interactive rhetorical partnership between researcher and participants in interpreting the role of new media and temporality in mobile crisis communication.

Conclusion

This chapter has highlighted the need for innovation in mobile crisis communication research and proposed a rhetorical approach to shed needed light on temporal and spatial nuances and complexities. Recent studies of map interpretation that rely on a socio-psychological orientation have produced mixed findings, with some studies showing clear benefits of map inclusion and others indicating uncertain outcomes. Understanding the reasons for these mixed findings is vital as the FCC is currently considering whether and how maps might accompany WEA messages in the future. Just as there is no way to render a map that will produce uniform interpretation and response (Cao et al., 2017), there is no singular WEA message that will ensure protective action taking. The rhetorical approach proposed here views maps and text as elements of rhetorical arguments. The arguments represented by WEA messages (with or without accompanying maps) compete with other arguments in a rhetorical process that may promote or impede convergence toward a desired conclusion. By using the interactive and informational capabilities of new media in more innovative research designs, scholars might help improve public understanding of timeliness and message specificity in mobile crisis communication.

Discussion Questions

1. WEA message templates and guidance are based on a semantic ideal of language use. That is, WEA messages are based on the premise that parsimonious and precise language use can overcome people's infinite capacity to interpret emergency messages in ways other than those intended by the message issuers. This semantic ideal leads to a concern for "optimized" messages, i.e., ones that will reduce "misinterpretation" and compel protective action. How might a rhetorical approach that emphasizes *interacting arguments* and *convergence* lead to different concerns in the design and use of mobile crisis communication messages?
2. Given that the meaning(s) of a map can never be totally fixed, what criteria should be considered when authorities are deciding whether or not to include a map image or a link to a map in a WEA message or other type of mobile crisis communication message?
3. How might a rhetorical approach, especially the concepts of *kairotopos* and "argument clusters" open up new possibilities for mobile crisis communication research?

References

Adams, B. (1988). Social versus natural time: A traditional distinction re-examined. In M. Young & T. Shuller (Eds.), *The rhythms of society* (pp. 198–226). London: Routledge.

Anthony, K. E., & Sellnow, T. L. (2011). Information acquisition, perception preference, and convergence by Gulf Coast residents in the aftermath of the Hurricane Katrina crisis. *Argumentation and Advocacy, 48*, 81–96. https://doi.org/10.1080/00028533. 2011.11821756

Anthony, K. E., Sellnow, T. L., & Millner, A. G. (2013). Message convergence as a message-centered approach to analyzing and improving risk communication. *Journal of Applied Communication Research, 41*, 346–364. https://doi.org/10.1080/00909882. 2013.844346

Bean, H., Liu, B. F., Madden, S., Sutton, J., Wood, M. M., & Mileti, D. S. (2016). Disaster warnings in your pocket: How audiences interpret mobile alerts for an unfamiliar hazard. *Journal of Contingencies and Crisis Management, 24*(3), 136–147. https://doi.org/10.1111/1468-5973.12108

Bean, H., Sutton, J., Liu, B. F., Madden, S., Wood, M. M., & Mileti, D. (2015). The study of mobile public warning messages: A research review and agenda. *Review of Communication, 15*, 60–80. https://doi.org/10.1080/15358593.2015.1014402

Bean, H., Wood, M. M., Mileti, D., Liu, B. F., Sutton, J., & Madden, S. (2014). *Comprehensive testing of imminent threat public messages for mobile devices.* First Responders Group, U.S. Department of Homeland Security. Retrieved from: www.nccpsafety. org/assets/files/library/Imminent_Threat_Public_Messages_for_Mobile.pdf

Burke, K. (1945). *A grammar of motives.* New York, NY: Prentice Hall.

Cao, Y., Boruff, B. J., & McNeill, I. M. (2016). Is a picture worth a thousand words? Evaluating the effectiveness of maps for delivering wildfire warning information.

International Journal of Disaster Risk Reduction, 19, 179–196. https://doi.org/10.1016/j.ijdrr.2016.08.012

Cao, Y., Boruff, B. J., & McNeill, I. M. (2017). The smoke is rising but where is the fire? Exploring effective online map design for wildfire warnings. *Natural Hazards, 88*(3), 1473–1501. https://link.springer.com/article/10.1007/s11069-017-2929-9

Casteel, M. A., & Downing, J. R. (2016). Assessing risk following a wireless emergency alert: Are 90 characters enough? *Homeland Security & Emergency Management, 13,* 95–112. https://doi.org/10.1515/jhsem-2015-0024

Castells, M. (2000). *The rise of the network society: The information age: Economy, society, and culture* (Vol. 1, 2nd ed.). Malden, MA: Blackwell Publishers.

Charmaz, K. (2003). Grounded theory. In J. Smith (Ed.), *Qualitative psychology: A practical guide to research methods* (pp. 81–110). Thousand Oaks, CA: Sage.

Coombs, W. T. (2010). Crisis communication and its allied fields. In W. T. Coombs & S. J. Holladay (Eds.), *The handbook of crisis communication* (pp. 54–64). Hoboken, NJ: Wiley.

Crowe, A. (2011). The social media manifesto: A comprehensive review of the impact of social media on emergency management. *Journal of Business Continuity & Emergency Planning, 5,* 409–420. www.ncbi.nlm.nih.gov/pubmed/21482509

Dransch, D., Rotzoll, H., & Poser, K. (2010). The contribution of maps to the challenges of risk communications to the public. *International Journal of Digital Earth, 3,* 292–311. https://doi.org/10.1080/17538941003774668

Duncheon, J. C., & Tierney, W. G. (2013). Changing conceptions of time: Implications for educational research. *Review of Educational Research, 83,* 236–272. https://doi.org/10.3102/0034654313478492

Flaherty, M. G. (2011). *The texture of time: Agency and temporal experience.* Philadelphia, PA: Temple University Press.

Fuller, T., & Pérez-Peña, R. (2017, October 13). In California, fires so fast hesitation proved lethal. *New York Times.* Retrieved from: www.nytimes.com/2017/10/13/us/california-wildfires-victims.html

Hagemeier-Klose, M., & Wagner, K. (2009). Evaluation of flood hazard maps in print and web mapping services as information tools in flood risk communication. *Natural Hazards and Earth System Sciences, 9,* 563–574. https://doi.org/10.5194/nhess-9-563-2009

Hariman, R. (1989). Time and the reconstitution of gradualism in King's address: A response to Cox. In M. C. Leff & F. J. Kauffeld (Eds.), *Texts in context: Critical dialogues on significant episodes in American political rhetoric* (pp. 205–217). New York, NY: Routledge.

Hassard, J. (Ed.). (1990). *The sociology of time.* London: Macmillan.

Haynes, K., Barclay, J., & Pidgeon, N. (2007). Volcanic hazard communication using maps: An evaluation of their effectiveness, *Bulletin of Volcanology, 70,* 123–138. https://doi.org/10.1007/s00445-007-0124-7

Jacques, E. (1982). *The form of time.* New York, NY: Crane Russak.

Lee, H., & Liebenau, J. (2000). Time and the internet at the turn of the millennium. *Time & Society, 9,* 43–56. https://doi.org/10.1177/0961463X00009001003

Liao, T. F., Beckman, J., Marzolph, E., Riederer, C., Sayler, J., & Schmelkin, L. (2013). The social definition of time for university students. *Time & Society, 22,* 119–151. https://doi.org/10.1177/0961463X11404385

Liu, B. F., Wood, M. M., Egnoto, M., Bean, H., Sutton, J., Mileti, D., & Madden, S. (2017). Is a picture worth a thousand words? The effects of maps and warning messages on how publics respond to disaster information. *Public Relations Review, 43*, 493–506. https://doi.org/10.1016/j.pubrev.2017.04.004

Madden, S. (2017). The clock is ticking: Temporal dynamics of campus emergency notifications. *Journal of Contingencies & Crisis Management, 25*, 370–375. https://doi.org/10.1111/1468-5973.12162

McKerrow, R. E. (1990). The centrality of justification: Principles of warranted assertability. In D. C. Williams & M. D. Hazen (Eds.), *Argument theory and the rhetoric of assent* (pp. 17–32). Tuscaloosa, AL: University of Alabama Press.

Middleton, M. K., Senda-Cook, S., & Endres, D. (2011). Articulating rhetorical field methods: Challenges and tensions. *Western Journal of Communication, 75*, 386–406. https://doi.org/10.1080/10570314.2011.586969

Mileti, D. S., & Sorensen, J. H. (1990). *Communication of emergency public warnings: A social science perspective and state-of-the-art assessment.* Oakridge, TN: Oak Ridge National Laboratory.

Moore, L. K. (2011). *Emergency Alert System (EAS) and all-hazard warnings.* Darby, PA: Diane Publishing.

Nowotny, H. (1994). *Time: The modern and postmodern experience.* Cambridge, UK: Polity Press.

Palomino, J., & Veklerov, K. (2017, October 12). Sonoma county officials opted not to send mass alert on deadly fire. *SFGate.* Retrieved from: www.sfgate.com/bayarea/article/Sonoma-County-officials-opted-not-to-send-mass-12271773.php

Perelman, C., & Olbrechts-Tyteca, L. (1969). *The new rhetoric: A treatise on argumentation.* Notre Dame, IN: University of Notre Dame Press.

Rämö, H. (1999). An Aristotelian human time-space manifold: From chronochora to kairotopos. *Time & Society, 8*, 309–328. https://doi.org/10.1177/0961463X99008002006

Romm, T. (2017, August 26). Before Hurricane Harvey, wireless carriers lobbied against upgrades to a national emergency alert system. *Recode.* Retrieved from: www.recode.net/2017/8/26/16205010/hurricane-harvey-texas-att-verizon-wireless-telecom-emergency-alert-system

Saulmon, G. (2013, July 1). At Sports World in East Windsor, camp director describes moments before—and after—the dome blew off. *MassLive.* Retrieved from: www.masslive.com/news/index.ssf/2013/07/at_sports_world_in_east_windso.html

Scott, J. B. (2006). Kairos as indeterminate risk management: The pharmaceutical industry's response to bioterrorism. *Quarterly Journal of Speech, 92*, 115–143. https://doi.org/10.1080/00335630600816938

Sellnow, T. L., & Seeger, M. W. (2013). *Theorizing crisis communication.* Hoboken, NJ: John Wiley & Sons.

Sellnow, T. L., Ulmer, R. R., Seeger, M.W., & Littlefield, R. (2009). *Effective risk communication: A message-centered approach.* New York, NY: Springer.

Smith, J. E. (1969). Time, times, and the "right time"; *chronos* and *kairos*. *The Monist, 53*(1), 1–13. https://doi.org/10.5840/monist196953115

St. John, P. (2017, December 29). Alarming failures left many in path of California wildfires vulnerable and without warning. *Los Angeles Times.* Retrieved from: www.latimes.com/local/lanow/la-me-fire-warnings-failure-20171229-story.html

Stephens, K. K., Ford, J. L., Barrett, A., & Mahometa, M. J. (2014). Alert networks of ICTs and sources in campus emergencies. In S. R. Hiltz, M. S. Pfaff, L. Plotnick & A. C. Robinson (Eds.), *Proceedings of the 11th International ISCRAM Conference* (pp. 650–659). University Park, PA: ISCRAM. Retrieved from: www.researchgate.

net/profile/Keri_Stephens/publication/262876448_Alert_Networks_of_ICTs_ and_Sources_in_Campus_Emergencies/links/53d704b60cf228d363eabf87.pdf

Venette, S. J. (2003). *Risk communication in a high reliability organization: APHIS PPQ's inclusion of risk in decision making.* Doctoral dissertation. North Dakota State University, Fargo.

Wajcman, J. (2008). Life in the fast lane? Towards a sociology of technology and time. *The British Journal of Sociology, 59*(1), 59–77. https://doi.org/10.1111/ j.1468-4446.2007.00182.x

Wood, M. M., Bean, H., Liu, B. F., & Boyd, M. (2015). *Comprehensive testing of imminent threat public messages for mobile devices: Updated findings.* First Responders Group, U.S. Department of Homeland Security. Retrieved from: www.dhs.gov/publication/ wea-comprehensive-testing-imminent-threat-public-messages-mobile-devices-updated

Wood, M. M., Mileti, D. S., Kano, M., Kelley, M. M., Regan, R., & Bourque, L. B. (2012). Communicating actionable risk for terrorism and other hazards. *Risk Analysis, 32*, 601–615. https://doi.org/10.1111/j.1539-6924.2011.01645.x

Wood, M. M., Mileti, D. S., Bean, H., Liu, B. F., Sutton, J., & Madden, S. (2017). Milling and public warnings. *Environment and Behavior.* Online first. https://doi. org/10.1177/0013916517709561

Zerubavel, E. (1982). The standardization of time: A sociohistorical perspective. *American Journal of Sociology, 88*(1), 1–12. www.jstor.org/stable/2779401

7

TRANSPORTATION NETWORK ISSUES IN EVACUATIONS

Tarun Rambha, Ehsan Jafari, and Stephen D. Boyles

Natural or human-made disasters often force individuals to evacuate to safer places. Depending on the nature of the disaster, evacuations may require people to travel short distances (as in the case of building fires, earthquakes, etc.) or long distances (as in the case of hurricanes, forest fires, volcanoes, hazmat leaks, etc.), and may affect a city, multiple states, or at times an entire nation. While small-scale evacuations do not affect transportation networks significantly, it is sometimes challenging to get first responders to the evacuating area during the response stage. Mock drills are helpful in preparing for this type of evacuation.

In contrast, evacuations that involve moving people across cities or states place an enormous amount of strain on infrastructure and financial resources. Such events are riddled with uncertainty since the disaster usually lasts for several hours or days. During this period, the strength of the disaster and geographical extent of affected area changes over time and are impossible to predict accurately, making it difficult for individuals and governments to react optimally. For instance, if a hurricane evacuation order is given early, individuals will have sufficient time to evacuate, but if the hurricane changes course, the evacuation may have been unnecessary, wasting effort and disrupting lives. On the other hand, delayed evacuation orders may put lives at risk, and may not allow enough time for safe evacuation.

Table 7.1 lists some of major natural disasters that have resulted in evacuations of approximately at least a million residents. Incidentally, all of them occurred in the past two decades, which is probably due to rapid urban development and population increase. Coupled with drastic changes in global climate, evacuations are expected to become more difficult in the future (Barrett, Ran, & Pillai, 2000). While most large-scale evacuations were triggered by hurricanes

TABLE 7.1 Disasters resulting in mass evacuations

Year	Name	Location	Evacuees (in millions)
1998	Yangtze River floods	China	14
1999	Hurricane Floyd	U.S. and Caribbean Islands	2.6
2005	Hurricane Rita	U.S. and Caribbean Islands	3
2007	Typhoon Krosa	China and Taiwan	1.4
2007	California wildfires	U.S.	1
2008	Hurricane Gustav	U.S. and Caribbean Islands	1.9
2013	Cyclone Phailin	India	1
2014	Typhoon Hagupit	Philippines	1
2016	Hurricane Mathew	U.S. and Caribbean Islands	2.5
2017	Hurricane Irma	U.S. and Caribbean Islands	7

and flooding, several other past instances have necessitated challenging evacuations due to their uniqueness, geographical location, and other constraints. Examples include the 1941 evacuation of Moscow (USSR) during WWII, the 2001 World Trade Center attack (U.S.), the 2002 evacuation of Goma (Democratic Republic of Congo) after the eruption of Mount Nyiragongo, and the 1986 Chernobyl (Ukraine) and 2011 Fukushima (Japan) nuclear disasters. Disasters can also have cascading effects. For example, the Fukushima nuclear disaster was due to a tsunami that was set off by an earthquake in the Pacific (Holguín-Veras et al., 2014). The rare nature of these events makes it all the more important to study hypothetical scenarios surrounding critical infrastructure which can improve preparedness for mass evacuations.

From a transportation perspective, resource constraints such as limited roadway capacity and vehicle availability restrict the ability to plan a safe and efficient evacuation. A major challenge in such scenarios is deciding when to evacuate, and how to efficiently handle infrastructure and human resources originally designed for regular operations. In the sections that follow, we discuss the difficulties that surround a large-scale evacuation and present a summary of studies that use mathematical models to address these problems. We conclude by also highlighting some of the challenges in post-disaster management.

Transportation Issues

A lack of adequate resources poses challenges for evacuation management. Particularly in the context of transporting people, roadway capacity is far less than demand. Roadway infrastructure is sized for typical demand loads, and not rare mass evacuation events. Many disasters such as earthquakes and flash floods can also render roads unusable. Further, uncoordinated departures and actions of self-interested agents compound this problem and result in massive traffic jams. For example, during Hurricane Rita, more than three million people evacuated the Houston area. The resulting traffic jams spanned over a hundred miles and caused gridlocks that took over a day to clear. Most individuals spent 12–20 hours on the road to reach the nearby cities of San Antonio and Austin. A heatwave during the same time resulted in nearly a hundred deaths, due to exhaustion, among those who were evacuating (Soika, 2006).

Vehicle availability also plays a major role in evacuation, even in regions with a high per-capita automobile ownership rate, since households without automobiles often have fewer resources in general. Limited vehicle availability also needs to be factored in while evacuating vulnerable populations such as those in hospitals, assisted living facilities, and prisons. The absence of a personal vehicle results in dependence on mass transportation, or reliance on friends or family with vehicles.

For instance, over 30 percent of the households in major East Coast cities in the U.S. do not own a vehicle and in New York the percentage is as high as 56 percent (Sivak, 2013). Urbina and Wolshon (2003) report that about 25–30 percent of residents in New Orleans did not have a vehicle before Hurricane Katrina. In this context, city agencies not only have to create transit routes and schedules that meet the passenger demand but must also communicate this information effectively and in advance.

Evacuations also present a challenge to first respondents, as most parts of the region being evacuated may not be accessible. Such situations may occur due to the lack of emergency lanes in parts of the network, due to dysfunctional traffic control systems, and due to gridlock created by evacuees. For instance, during Hurricane Sandy, nearly 3,500 traffic signals stopped operating (Gibbs & Holloway, 2013). Lack of power and flooding of road networks caused due to flash floods and hurricanes is another important factor to take into account. Adding to these problems, the need for assistance goes up significantly during the event of a disaster. During Hurricane Sandy, New York City's 911 emergency received a record volume of nearly 20,000 calls per hour. As a result, many callers got a busy signal. To reach out for those in serious emergencies, police and fire department personnel had to use inflatable and jon boats (Gibbs & Holloway, 2013). Very often vulnerable populations and hospital patients trapped due to floods have to be airlifted (Gray & Hebert, 2007).

Evacuations can also cause serious second-order effects such as shortage of gasoline. Given the uncertainty surrounding an emergency, evacuees typically fill up their vehicles, which leads to a high degree of supply–demand imbalance. This issue is further compounded when the supply gets cut off because of the disaster. Gasoline shortages can influence the destinations of travelers or could alternately leave them stranded on freeways, forcing them to abandon their vehicles and find alternate means of evacuating. Although steps such as rationing have been taken in the past, as during Hurricane Sandy (Kaufman, Qing, Levenson, & Hanson, 2012) and Hurricane Matthew, it continues to be a major concern for policymakers handling evacuations.

Emergencies affect other critical infrastructure that can have impacts on transportation and travel choices. For instance, earthquakes, forest fires, and winds during hurricanes can disrupt power and water networks. Lack of power can hamper individuals' ability to communicate and receive information from TV, radio, and the Internet. With the rise of information and communication technologies (ICTs), social media applications such as Facebook and Twitter have helped send and receive information from mobile devices much faster than traditional radio/TV and face-to-face communication. These outlets also have the potential to validate information and reinforce the decision to evacuate (Stephens, Jafari, Boyles, Ford, & Zhu, 2015). However, power outages can severely affect these channels, and can cause confusion and result in poor evacuation-related choices. In the future, widespread adoption of electric and connected autonomous vehicles will present new challenges for evacuation in the absence of power.

From the standpoint of building these models, there are institutional challenges. The regions affected by disasters do not neatly align with political boundaries, and evacuations involve cross-jurisdictional coordination. The need for coordination among multiple cities, counties, or states creates difficulties in obtaining the necessary data in a common format, and in building models with a consistent level of detail across a region. Recent research in mega-regional network modeling (Jafari, Pandey, & Boyles, 2016; Yahia, Pandey, & Boyles, 2018) suggests that large networks can be partitioned into smaller networks (perhaps one per county or affected state), modified in a way that ensures consistency across the region—but such methods have not yet been tested in practice.

As outlined above, several transportation issues accompany large-scale evacuations. These problems may not be fully resolvable, but they can be handled better with proper planning. While it is not possible to conduct large-scale mock drills, one can simulate the effects of a disaster using mathematical models. Major topics of interest in the transportation network modeling community are centered around developing tools with a decision support component that can help government agencies and individuals make better choices before and during an evacuation. These problems are often demanding as they lie at the intersection of multiple disciplines such as behavioral and social sciences, engineering, and optimization. Furthermore, they involve working with limited data and a high degree of uncertainty, factors that make it important to design robust tools that address a wide range of real-world scenarios.

Most studies on evacuations are centered around hurricanes and a few are on wildfires because they are more common than other disasters. For example, an average of 1.7 hurricanes strike the U.S. coast each year (Landsea, 2015). Also, they offer a greater lead times compared to no-notice evacuations and hence can be managed better. In the following sections, we summarize previous research on choices individuals make in the context of evacuations and discuss evacuation planning and operational models.

Understanding Behavior: When, Where, and How People Evacuate

Managing a disaster efficiently necessitates a thorough understanding of human behavior in various emergency situations. A central problem in this context is to predict when and how individuals evacuate and where they evacuate to. The ability to model this behavior allows planners to simulate and optimize decisions such as estimating clearance times, providing optional and mandatory evacuation orders, designing shelter locations, activating contraflow lanes, suggesting routes, and creating emergency transit options. More details on these topics will be discussed in the next section.

Evacuation response begins with warnings and predictions that are issued by public agencies such as the National Hurricane Center and National Weather Service. These official warnings are general in nature and are not personalized. In the case of hurricanes, predictions are typically provided for the next 3–5 days using flooding and atmospheric models such as SLOSH (Jelesnianski, Chen, & Shaffer, 1992). Local governments can issue voluntary or mandatory evacuation orders based on this information and other regional constraints. Households, however, may decide to evacuate proactively even before warnings are issued (Dow & Cutter, 2000).

Several psychological studies attempt to understand the influence of the warning process (Sorensen, 2000). After hearing warnings, individuals typically understand, believe, personalize, and take action (Mileti & Sorensen, 1987; Mileti & Peek, 2000). The behavioral responses are surprisingly similar for most natural disaster warnings (Leach,1994). Frequency, source, and distribution medium of warnings can have a significant impact on risk perception among individuals (Mileti & O'Brien, 1992; Lindell, Lu, & Prater, 2005). Lindell and Perry (2003) suggest that, in addition, the wording and content can highlight the urgency of the situation and play a major role in the departure and destination choices of evacuees.

Several other factors can accentuate the effect of public warnings. Influence of friends and family and business closures can induce conformity effects and can make people perceive that the situation is dangerous and that they must leave (Baker, 1991). ICTs and social media aid in this process since they have started to be the most popular place to share information (Sutton, Palen, & Shklovski, 2008; Stephens, Jafari, Boyles, Ford, & Zhu, 2015). Arlikatti, Lindell, Prater, and Zhang (2006), however, warn that certain sections of society may not be adept at using these technologies, and hence traditional door-to-door communication may still be necessary. Recent studies have also explored the effect of a single individual's actions on his/her social circles and the cascading effects it can cause (Hasan & Ukkusuri, 2011).

Back-to-back disasters can also cause nontrivial reactions to warnings. For instance, massive evacuations during Hurricane Rita were primarily a result of the extent of damage done by Hurricane Katrina. Also called *shadow evacuations*, these evacuations can overload the transportation network and prevent those who really need to exit the critical area (Zeigler, Brunn, & Johnson, 1981). On the other hand, Dow and Cutter (1998) warn about the "crying wolf" phenomenon, when evacuation messages are given for events that turn out to be not dangerous. Such complacency effects were visible during Hurricane Sandy after the previous Hurricane Irene was not as threatening as it was expected to be.

When to Evacuate?

A crucial choice made by individuals affected by a disaster is the decision to evacuate or shelter-in-place. Further, if a household evacuates, predicting the departure time is essential for all transportation models as well as for identifying the kind of information to be provided to stage departures. This task is typically carried out using a one-stage or two-stage method (Pel, Bliemer, & Hoogendoorn, 2011; Murray-Tuite & Wolshon, 2013). Both approaches are random utility-based econometric models in which the disutility of choices are assumed to be a function (typically linear) of various sociodemographic and disaster-specific influencing attributes and an error term representing factors that are observable to the individuals but not the analyst. They require stated preference surveys, in which the interviewer presents hypothetical scenarios, or revealed preference data (actual observations during a disaster) to understand the influence of different attributes. Individuals are assumed to pick a choice that minimizes his/her disutility and, using data, the signs of the attributes and their statistical significance are established.

The key difference is that the one-stage model predicts both the decision to evacuate as well as the departure time. On the other hand, in the more commonly used two-stage method, the first stage is used to predict if the household evacuates and in the second stage the cumulative demand is spread over time using time-dependent participation rates or S-shaped patterns (FEMA, 1999). These S-shaped patterns were first observed during Hurricane Opal and other events in the past exhibited similar responses. Various S-shaped distributions such as Weibull (Lindell, 2008), Sigmoid (Kalafatas & Peeta, 2009; Xie, Lin, & Waller, 2010), and Rayleigh distribution (Tweedie, Rowland, Walsh, Rhoten, & Hagle, 1986) have been used in the literature.

Econometric models for the first stage differ in the assumptions made on the error terms or the unobserved attributes. More complicated error structures permit realistic representation of choices and yield better fit. However, they also turn out to be difficulty to estimate. Commonly used models are logit (Whitehead et al., 2000; Wilmot & Mei, 2004), mixed logit (Hasan Ukkusuri, Gladwin, & Murray-Tuite, 2011; Sarwar, Anastasopoulos, Ukkusuri, Murray-Tuite, & Mannering, 2016), and ordered probit models (Xu, Davidson,

Nozick, Wachtendorf, & DeYoung, 2016) were used to predict the probability with which an individual or a household evacuates. Participation rate curves have also been noted to be applicable to events that have sufficient a warning time such as wildfires (Wolshon & Marchive, 2007; Dennison, Cova, & Mortiz, 2007; Cova, Dennison, & Drews, 2011). Choice models have also been used to understand evacuation decisions during earthquakes and tsunamis (Troncoso Parady & Hato, 2016; Lindell et al., 2015; Wei et al., 2017).

A major drawback, however, with the two-stage method is that it is not endogenously determined by the threat (Murray-Tuite & Wolshon, 2013), and is sensitive to the choice of the distribution (Yazici & Ozbay, 2008) and the rate parameters (slopes of the S-shape) do not have a strong behavioral justification. Hence, it does not capture sudden changes in departure rates caused by time-of-day variations and issuance of evacuation orders (Lindell & Prater, 2007; Dixit, Montz, & Wolshon, 2011). Fu, Wilmot, Zhang, and Baker (2007) attempted to fix this issue using data from Hurricanes Floyd and Andrew but transferability to other scenarios is yet to be established.

Alternately, in a less frequently explored method, departure time choices can be jointly modeled along with the decision to evacuate or not. Fu and Wilmot (2004) used a repeated binary logit model for hurricane evacuation which is applied over six-hour time windows from data on Hurricane Andrew. Gudishala and Wilmot (2012) used a nested-logit model for addressing the same problem. While these methods either use prevailing conditions for predicting evacuation rates or assume that the future is fully known, a more realistic modeling assumption is to include future forecasts in the disutilities of individuals. Serulle and Cirillo (2017) use a dynamic discrete choice framework to capture these effects.

Table 7.2 lists various sociodemographic and disaster-specific factors that can potentially affect the propensity to evacuate. Several researchers have used qualitative and quantitative methods to analyze the significance of these factors from survey data. However, the results are very mixed. For instance, some studies find that income is not significant (Whitehead et al., 2000; Bateman & Edwards, 2002), while some found it to be significant and has a positive influence on evacuation (i.e., the higher the income, the greater are the odds of evacuating) (Gudishala & Wilmot, 2012; Elliott & Pais, 2006). A few have also observed a negative effect (Zhang, Prater, & Lindell, 2004; Smith & McCarty, 2009). The reason for such variation is because the models are fit to a specific data set. The choice of the model, data, and other unobserved factors can influence the significance and signs of different attributes. Hasan, Mesa-Arango, Ukkusuri, and Murray-Tuite (2012) fuse data from multiple evacuation events to improve the transferability of choice models. Other literature that analyses the effects of these variables can be found in Sorensen (2000) and Murray-Tuite, Yin, Ukkusuri, and Gladwin (2012). For an in-depth meta-analysis of attributes and their effects from 49 studies since 1991, refer to Huang, Lindell, M. K., and Prater (2016).

TABLE 7.2 Factors affecting the decision to evacuate

Sociodemographic	*Disaster-specific*
Age	Source of evacuation notice
Gender	Voluntary/mandatory notice
Ethnicity	Hurricane category (Saffir-Simpson scale)
Household type and location	Storm surge
Household composition	Projected trajectory
Number of vehicles	Wildland urban interface hazard scale
Education	Roentgen equivalent man
Occupation	Volcanic explosivity index
Income	Richter scale
Presence of pets	Integrated tsunami intensity scale
Disability	Precipitation amount
Residence period	Congestion levels
Experience evacuating	
Refuge location and proximity	

Where to Evacuate?

Common destination choices during emergencies include houses of friends and family, hotels, and shelter locations (Lindell, Kang, & Prater, 2011; Wu, Lindell, & Prater, 2012; Mesa-Arango, Hasan, Ukkusuri, & Murray-Tuite, 2013). After estimating the demand for evacuation, gravity-based or econometric models are used to predict what fraction of evacuees pick different locations and accommodation types. In a gravity-based trip distribution approach, the fraction of travelers evacuating to a certain destination is assumed to be proportional to the attraction potential of a location and inversely proportional to travel distance (impedance). The attraction potential depends on factors such as location (proximity to highways), accommodation type, and population in the region containing the destination (Whitehead et al., 2000; Brodie, Weltzien, Altman, Blendon, & Benson, 2006; Cheng, 2007). For instance, the attraction potentials for hotels depend on the availability of rooms and price (Cuellar, Kubicek, Hengartner, & Hansson, 2009). A study by Cheng, Wilmot, & Baker (2011) also considers dynamic impedance factors using predicted hurricane trajectories. Evacuees may also first choose proximate destinations, which are locations that fall outside the risk area and then proceed to a final destination (Lindell & Prater, 2007).

To specifically predict accommodation types, discrete choice models similar to those used to estimate demand are typically employed (Carnegie & Deka, 2010). These models have greater flexibility since they can identify the effects of various influencing attributes on the choice process. For instance, individuals with higher income and education tend to use hotels more than others, and use shelters less (Brodie et al., 2006). Additionally, factors such as age, education, household composition, ethnicity, pet ownership, type of evacuation, severity of disaster, and travel distance have been found to affect accommodation type (Mileti & O'Brien, 1992; Whitehead et al., 2000; Cheng, Wilmot, & Baker, 2008; Murray-Tuite et al., 2012). Recent studies by Golshani, Shabanpour, Auld, Mohammadian, and Ley (2018) and Yang, Morgul, Ozhay, and Xie (2016) focus on joint models for departure decisions and destination choices using structural equation modeling and joint discrete-continuous methods respectively.

How to Evacuate?

Once an individual or a household decides to head to a safe destination, the next major choice is to select a mode and a route for evacuation. In most cases, individuals simply choose to drive unless vehicle availability is an issue (Wu et al., 2012; Deka & Carnegie, 2010). The route selection process, on the other hand, is more complex. The aggregate effect of individuals' choices dictates congestion levels and the time required to clear traffic. Under normal conditions, when demand for traffic is available, planners can estimate the volumes, speeds, queue lengths, and travel times on different roadway links by making additional assumptions on the route choice behavior of travelers. One option is to assume that users are self-interested agents who try to minimize their travel time, which results in a flow pattern called *user equilibrium* (UE) (Wardrop, 1952). In this state, all used paths between an origin-destination pair have equal and minimal travel times. UE-based models further assume that travelers learn the optimal routes through experience. Alternately, if the agents in a network can be guided by a central system (using variable message signs or mobile apps), it is possible to minimize the total evacuation time. This flow pattern, called the *system optimum* (SO), is ideal for evacuations but is harder to enforce since it requires full compliance. It does, however, provide a benchmark for efficiently managing an evacuation. It could also be highly applicable in the future for a system with connected autonomous vehicles. An overview of mathematical formulations for route choice models can be found in Boyles and Waller (2011a).

Depending on the scale of the problem and the availability of data and computational resources, one can choose to study traffic and route choice behavior at different levels (Chen & Zhan, 2008). Static assignments of traffic to roadways do not explicitly model departure time choices but assume steady-state conditions and simplified delay equations, which makes them highly tractable. On the other hand, dynamic route choice models allow more granular representation of traffic and can mimic vital features such as queue formation and spillback (Peeta & Ziliaskopoulos, 2001; Chiu et al., 2011).

Travelers can also make decisions en route at each intersection based on the current and predicted network conditions that are relayed using advance traveler information systems and mobile phones (Unnikrishnan &

Waller, 2009; Pel, Bliemer, & Hoogendoorn, 2009; Gao, 2012; Rambha, Boyles, Unnikrishnan, & Stone, 2018). Surveys indicate that compliance to rerouting recommendations is higher during evacuations (Knoop, Hoogendoorn, & van Zuylen, 2010; Robinson & Khattak, 2010). These types of hybrid routing models are suited for scenarios in which travelers do not have full information and also do not have the option of learning optimal routes through experience. Furthermore, during evacuations, travel demand is highly unpredictable and other dynamic congestion mitigation measures such as contraflow may be in place. Some researchers have also attempted to predict the probability with which travelers take familiar routes, recommended routes, change routes midway, etc., using revealed route choice surveys and econometric models instead of equilibrium-based approaches (Pel et al., 2008; Sadri, Ukkusuri, Murray-Tuite, & Gladwin, 2014). While most existing studies analyze demand and supply components of travel separately, a recent study by Ukkusuri et al. (2017) provided a tool that integrates these two problems and suggests predicts congestion during evacuations. A list of some traffic assignment models developed for evacuations is shown in Table 7.3.

Individuals also prefer to use freeways, familiar routes, and regions with gas stations and better cellphone connectivity (Akbarzadeh & Wilmot, 2014). This creates more traffic along some portions of the network, despite having alternate uncongested options (Dow & Cutter, 2002). Another factor to consider during modeling route choice is background traffic that is comprised of vehicles with far-off origins and destinations, which pass through the regions being evacuated (Brown, White, van Slyke, & Benson, 2009; Zheng, Chiu, Mirchandani, & Hickman, 2010). Also, individuals may prefer to travel

TABLE 7.3 Features of evacuation route choice models

Model	UE	SO	Other	Static	Dynamic
CEMPS (Pidd, de Silva, & Eglese, 1993, 1996)		X			X
NETVAC (Sheffi, Mahmassani, & Powell, 1982)		X			X
MASSVAC (Hobeika & Kim, 1998)	X			X	
TEDSS (Sherali, Carter, & Hobeika, 1991)		X		X	
OREMS (Rathi & Solanki, 1993; Li, Yang, & Wei, 2006)		X			X
EVAQ (Pel, Bliemer, & Hoogendoorn, 2008; Pel, Hoogendoorn & Bliemer, 2010)			X		X
DYNASMART (Han, Yuan, Chin, & Hwang, 2006)	X				X
DynusT (Chiu & Mirchandani, 2008; Ayfadopoulou, Stamos, Mitsakis, & Grau, 2012)		X	X		X
A-RESCUE (Ukkusuri et al., 2017)			X		X

as a unit with friends and families who live nearby. Such meeting point-type features have also been included in some past studies (Murray-Tuite & Mahmassani, 2003, 2004; Lin, Eluru, Waller, & Bhat, 2009).

Tools for Planning and Carrying Out an Evacuation

Building on top of the behavioral models discussed in the previous section, various planning and operational problems can be formulated. One can think of these as "what if" scenarios that are simulated and the best among the different options can be used in practice. These problems have consequential policy implications and solving them can greatly improve the evacuation process.

Issuance of Evacuation Orders

Among the factors that influence the decision to evacuate, voluntary or mandatory evacuation orders have a significant impact on the demand for roads, and consequently on roadway congestion. Thus, by issuing such evacuation orders in a staggered manner, it is possible to distribute the demand and avoid shadow evacuations. Such staged departures can reduce the overall time required to evacuate and also adapt to the changing dynamics of the disaster. Several studies in the past assume full compliance and employ traffic simulators (such as the cell transmission model, CTM) to find departure rates and routes that optimize the clearance time (Yuan, Han, Chin, & Hwang, 2006; Villalobos, Chiu, Zheng, & Gautam, 2006; Dixit & Radwan, 2009; Bish & Sherali, 2013). Almost, all of them establish that phased evacuations can reduce traffic when compared with simultaneous departures. In other works, Cheng et al. (2008) study the effects of network topologies and population densities on evacuation times. Wolshon, Zhang, Parr, Mitchell, & Pardue (2015) evaluated the effect of phased departure profiles using an elaborate agent-based model. However, from a practical perspective, departures cannot be controlled at such fine resolutions but can be indirectly influenced using evacuation orders.

In order to effectively communicate evacuation orders, it is necessary to divide critical regions into different zones in advance based on potential risk. Zones must be few in number and easily identifiable using an alphabetical or a numbering scheme and all individuals must be informed of the zone they live in. The selection of critical regions can be done using risk maps created from topology, engineering judgment, and historic data (Arlikatti et al., 2006; Zhang et al., 2004). Zoning should make use of ZIP codes and landmarks to make it easily identifiable (Wilmot & Meduri, 2005). For example, following Hurricane Sandy, regions along the coast in New York City were divided into six zones numbered 1 to 6 (1 represents the region with highest risk and 6 is the region with lowest risk). Similarly, the Houston-Galveston area is divided into four zones: coastal, A, B, and C. The issue of zoning can also be formulated as an clustering problem by considering the impact of zoning on the clearance times (Hsu & Peeta, 2014).

Once evacuation zones are created, it is necessary to find an ideal sequence in which orders are issued. Demand estimation methods discussed earlier can be used to infer the compliance to such orders and roadway traffic can be predicted, which can be used as an objective for finding the optimal sequence. However, research on this topic is rather limited. Recently, Yi, Nozick, Davidson, Blanton, and Colle (2017) proposed a multistage model that addresses this problem. Information on current and predicted hurricane trajectories are first used to construct a scenario tree with a finite number of possibilities. The tree is then used to formulate a stochastic optimization program that recommends evacuation orders for different zones in different time periods for different future scenarios.

Contraflow, Signal Control, and Network Design

Experiences with major evacuation events indicate the need for additional capacity to manage the evacuation demand (Litman, 2006). Contraflow, defined as reversing a subset of links during the course of evacuation, has been suggested as an effective strategy to increase the network capacity and to minimize the evacuation time (Wolshon, 2001; Wolshon, Lefate, Naghawi, Montz, & Dixit, 2005). The idea of reversing the lanes to increase road capacity was first proposed by the Federal Emergency Management Agency (FEMA). The core problem of contraflow evacuation planning is selecting a subset of links to reverse and routing the traffic through the network. Experienced-based contraflow plans are usually not capable of finding critical road segments and may even worsen the situation.

General contraflow design problem is NP-hard (Rebennack et al., 2010).[1] For the case of single-source, single-sink problems, however, polynomial time algorithms have been proposed (Rebennack, Arulselvan, Elefteriadou, & Pardalos, 2010). Tuydes and Ziliaskopoulos (2004) used a UE-based model to solve the contraflow network problem. The problem admits an exponential number of feasible options and hence does not scale well to large problem instances. Hence, tabu-based search methods (Tuydes & Ziliaskopoulos, 2006) and other heuristics such as those that increase the network capacity by reversing the links with highest volume to capacity ratios (Kim, Shekhar, & Min, 2008) have been proposed. The experiments in the latter paper showed a reduction of more than 40 percent in evacuation time. Kalafatas and Peeta (2009) developed an integer programming formulation that also addresses the uncertainty in evacuation parameters such as population size and demand distribution, subject to a budget constraint. Their findings showed that the improvement in evacuation time is marginal after a budget threshold and that the lowest evacuation time is obtained when demand is uniformly distributed between different origin-destination pairs. Chiu and Mirchandani (2008) showed that contraflow strategy integrated with phase-based evacuation, distributing the evacuation demand in time, would significantly increase the performance of the evacuation plan. Dynamic lane reversal, altering lane directions in response to traffic conditions monitored by the sensors, was proposed by Hausknecht, Au, Stone, Fajardo, and Waller (2011) to enhance the traditional static lane reversal strategies. In their experiments, dynamic contraflow plan improved the traffic efficiency by 72 percent. To enhance our understanding from mass evacuation and the performance of contraflow evacuation plans in practice, researchers have mainly relied on simulation-base analysis. While microscopic traffic simulation models are very valuable to study the small-scale evacuation scenarios (Theodoulou &Wolshon, 2004; Lim & Wolshon, 2005; Yu, Pande, Nezamuddin, Dixit, & Edwards, 2014), macroscopic and agent-based traffic models have been employed to simulate large-scale emergency evacuation scenarios (Kirschenbaum, 1992; Kim et al., 2008; Chiu & Mirchandani, 2008; Naghawi & Wolshon, 2012; Wolshon et al., 2015).

As another major source of traffic delay, controls at intersections should also be configured to maximize the safety and throughput while practicing the contraflow operation. To address this problem, Cova and Johnson (2003) suggested an optimal lane-based routing algorithm (intersection crossing elimination) to reduce delays due to traffic signals and stop signs by prohibiting some turning movements. Xie and Turnquist (2009) also proposed extensions in which both the lane reversal and intersection crossing elimination components are jointly modeled using optimization techniques and heuristics. Xie et al. (2010) formulated the emergency evacuation plan with lane reversal and intersection conflict elimination as a bi-level network optimization problem. The upper-level problem optimizes the evacuation performance while the lower-level problem implements the CTM to simulate the network condition. Signals at intersections are also a major source of concern, since they are designed for regular traffic conditions. This can lead to gridlock, especially during the evacuation of dense urban areas. Research on this topic is sparse. Chen, Chen & Miller-Hooks (2007) analyze the impact of different signal plans for a hypothetical evacuation of Washington, DC. Hamza-Lup, Hua, Le, and Peng (2008) devise adaptive control algorithms that optimize the throughput. Noncompliance of drivers with traffic signals may be another issue during disasters. The information used to change the timing of traffic signals can also be provided directly to evacuees, informing them where congestion lies, and providing opportunities for taking alternate routes (Boyles & Waller, 2011b).

For disasters such as earthquakes or floods, in which the road network may be adversely affected, evacuation plans based on the network capacity may fail. Evacuation network design problem, defined as a tool for identifying the best candidate road segments to be constructed and improving critical road network segments have been proposed to maximize the performance of the evacuation plan. Hadas and Laor (2013) developed a model focusing on minimizing both the construction cost and evacuation time. The authors developed a heuristic to solve the problem for practical applications. The evacuation network design problem formulated in Nahum and Hadas (2017) minimizes the construction cost considering the current capacity of the network, infrastructure vulnerability, and demand stochasticity. A detailed review of proposed network design strategies for the evacuation purpose is presented in Abdelgawad and Abdulhai (2009).

Designing Shelter Locations

As highlighted earlier, evacuees typically head to houses of friends and family, hotels, and shelter locations. Among these choices, city and regional planners can make long-term decisions only on the locations and capacities of safe shelters. These decisions are crucial to an evacuation process since they can be used to indirectly influence individuals' choices pertaining to when and where to evacuate, and thus improve evacuation costs and times. They are also extremely important in small islands (especially in the Atlantic), since evacuees do not have a choice to travel large distances from impending disasters. The general approach to designing shelters is to formulate a Stackelberg game or a bi-level optimization problem, in which the planning decisions associated with the location and/or capacity of shelters are made at the upper level. In response, drivers select routes to get to the shelters in the lower level, which can be modeled using the route choice models discussed earlier. The goal is to thus find the best solution to the upper level, assuming that the drivers will respond optimally in the lower level. Most studies in the literature adhere to this framework, but they do differ in objective functions and the underlying assumptions in the second level. For example, Sherali et al. (1991) and Yazici and Ozbay (2007) assume that drivers in the lower level will follow an SO assignment of flow, while Ng, Park, and Waller (2010) suppose that a UE state is more likely. Features such as multiple objectives with risk measures, distance, time, etc. (Alçada-Almeida, Tralhão, Santos, & Coutinho-Rodrigues, 2009), stochasticity in destination choices (Kongsomsaksakul, Yang, & Chen, 2005), robust designs under demand uncertainty (Kulshrestha, Wu, Lou, & Yin, 2011), optimization of locations with limits on detour time in SO routing (Bayram, Tansel, & Yaman, 2015), and probabilistic models for link failures (Yazici & Ozbay, 2007) have also been incorporated.

For hurricane evacuations, stochasticity in track and intensity can be used to develop resource allocation strategies that are adaptive. Li, Xu, Nozick, and Davidson (2011) and Li, Nozick, Xu, and Davidson (2012) proposed a scenario-based optimization model which takes into account the set of predicted hurricane trajectories and computes the optimal subset of shelters to be opened under budgetary and staffing constraints while assuming that users route choices satisfy stochastic and dynamic user equilibrium conditions respectively. Li, Jin, and Zhang (2011) used a two-stage stochastic optimization model in which the first-stage decision variables are used to plan the locations and capacities of shelters, and the second-stage decision variables are related to transportation of evacuees and essential supplies from distribution centers. The objective included fixed costs for building permanent shelters and operating costs for transporting and housing evacuees.

For other types of disasters, Coutinho-Rodrigues, Tralhão, and Alçada-Almeida (2012) and Kılcı, Kara, and Bozkaya (2015) address the problem of locating shelters in earthquake-prone areas by maximizing a score that is a function of several parameters such as access to healthcare and electricity, type of terrain, and presence of sanitary systems. They also require the utilization of each opened shelter to be above a predetermined threshold. Alçada-Almeida et al. (2009) and Coutinho-Rodrigues et al. (2012) use a multi-objective optimization approach to determine where to open a fixed number of shelters in the event of a fire. Their objectives include distance to shelter, fire risk, and time to travel from shelters to a hospital.

Planning for Shared Modes of Transportation

As noted previously, when planning for evacuation, it is also necessary to address the needs of vulnerable populations and individuals/households without a vehicle. To this end, transportation planners must design public transit services that operate between stops and shelters specifically for evacuation purposes while minimizing operating costs (An, Cui, Li, & Ouyang, 2013) or evacuation time (Sayyady & Eksioglu, 2010). Transit systems can be broadly classified as fixed-route and demand-responsive systems.

Following Hurricane Katrina, many coastal cities in the U.S. created fixed-route transit plans (Wolshon et al., 2009). To predict the robustness of such fixed-route plans under potential hurricane trajectories, Naghawi and Wolshon (2010, 2012) used TRANSIMS, an agent-based model to simulate the impacts of different demand-loading patterns and schedules on evacuation times and congestion levels. They found that routing buses on arterials during off-peak periods was ideal since it reduced the time taken to clear auto traffic as well. On the other hand, routing buses on freeways was found to increase the time required for evacuation. Providing real-time information to evacuees using these services is important, since these fixed routes may differ from typical transit route configurations. Rambha, Boyles, and Waller (2016) describe how online route guidance systems can reduce travel delays when travelers are unaware of the exact location and arrival times of buses.

For demand-responsive systems, two problems are of particular interest. First, transit stops must be designed in near real-time based on the evacuation demand. Second, vehicles must be routed from these stops to shelter locations. For the first task, one may consider the census data from different regions, data on vehicle ownership, and distance constraints to estimate potential pickup points (Bian & Wilmot, 2018). Liu, Murray-Tuite, and Schweitzer (2011) use a mathematical program to find stops where children and evacuees could be relocated and picked up in case of an emergency. Once, the demand of evacuees is known, a classical combination optimization problem called the vehicle routing problem (VRP) can be used to route the buses and assign evacuees to different buses (Rui, Shiwei, & Zhang, 2009; Abdelgawad, Abdulhai, & Wahba, 2010; Zheng, 2014). These problems are also NP-hard and are notoriously difficult to solve. Generally, heuristics such as cluster-first route-second, genetic algorithms, hill-climbing methods, tabu search, and simulated annealing are used to solve such problems (Toth & Vigo, 2002). Hence, most studies propose heuristics that provide near-optimal solutions. The objectives of these problems could be to minimize the waiting time as well as the total travel time. The constraints of the model ensure that all evacuees are served, and it is possible to impose capacity constraints and limits on the time windows for picking up evacuees. Multi-trip versions, in which the buses make several trips between pick-up locations and shelters have also been formulated to address the constraints on capacity and the number of vehicles (Bish, 2011). Mathematical models in which the pickup points and schedules are jointly optimized under demand uncertainty have also been proposed (He, Zhang, Song, Wen, & Wu, 2009; Kulshrestha, Lou, & Yin, 2014; Hua, Dai, Wang, & Li, 2016; Swamy, Kang, Batta, & Chung, 2017). Chen and Chou (2009) also suggest performing transit routing along with contraflow using a bi-level optimization model.

VRP-type models have been formulated for wildfire evacuations as well (Shahparvari, Chhetri, Abbasi, & Abareshi, 2016; Shahparvari & Abbasi, 2017). Bish, Agca, and Glick (2014) also studied the problem of minimizing patients' risk while evacuating a hospital with constraints on ambulance and patient types. The advent of automated vehicles also presents new opportunities for shared mobility services during evacuations (see, for instance, Levin & Boyles, 2015), particularly for populations with limited private vehicle ownership.

Recovery

While there have been quite a few studies that focus on evacuation, the topic of recovery has received relatively little attention. A recent study by Donovan and Work (2017) found that travel times after Hurricane Sandy were higher than those during the evacuation process, highlighting the need for adequate planning for the recovery process. Although the transportation problems post-evacuation are similar to those during pre-evacuation, travelers are usually no longer at risk and hence the objectives of public agencies during this phase can be slightly different. Staged restoration policies can significantly reduce societal costs and inconvenience experienced by individuals. In this context, it is essential to measure the resiliency of networks to aid the recovery process. For instance, Chang and Nojima (2001) proposed two simple measures for measuring rail and roadway disruptions after the 1995 Hyogoken-Nanbu (Japan) earthquake. These include the ratio of the length of the network that is open to traffic to the length of network before the disaster and an accessibility measure that captures the extra distance one must travel post-disaster. Bruneau et al. (2003) and Zhu, Ozbay, Xie, and Yang (2016) suggested that the area of triangular portions in plots between quality of infrastructure and time after a disaster can be used a metric for comparing different strategies. Zhu et al. (2016) and Mudigonda, Ozbay, and Bartin (2018) analyzed the number of taxi trips across time in NYC and the number of days spent in restoration of transit and rail services in New Jersey respectively after Hurricane Sandy. While these studies have tried to quantify recovery activities that were used in the past, a few studies have addressed the problem of finding the optimal restoration strategies. For instance, Chen and Miller-Hooks (2012) developed a stochastic optimization model for transporting essential supplies while also selecting recovery actions for fixing roadway links. The cost and duration of reconstruction activities were also taken into account. Çelik, Ergun, and Keskinocak (2015) suggested methods to find the optimal plans for clearing roads of debris that accumulated during earthquakes and storms.

Other major transportation challenges during post-disaster period are associated humanitarian aid distribution. Disaster such as earthquakes and hurricanes result in an increased demand for food, clothing, medical supplies, etc. (Holguín-Veras & Jaller, 2011). Further, organization and transportation of relief workers is also a difficult issue, especially in the context of disasters occurring in regions that are not easily accessible, such as Haiti and Nepal. Duran, Ergun, Keskinocak, and Swann (2013) highlight the need for pre-positioning supplies that are crucial for the recovery phase. Classic optimization models such as the p-median and covering problems can be formulated to find warehouse locations that serve a large fraction of the demand at low costs and within short duration. They can also be modified to jointly find the ideal network rehabilitation schemes (see Liberatore, Ortuño, Tirado, Vitoriano, & Scaparra, 2014, for example). Such problems are broadly studied under the moniker of humanitarian logistics. A broad overview of this area of research can be found in Kovács and Spens (2007), Overstreet, Hall, Hanna, and Rainer (2011), and Çelik et al. (2012).

Conclusion

Large-scale evacuations impose unique demands on transportation systems, in terms of the number of travelers making use of the system, their spatial travel patterns (origins and destinations), and the timing of their trips.

Transportation infrastructure is not designed for exceptional circumstances like evacuations, and traditional modeling approaches often fall short of the needs of effective emergency planning. To support disaster logistics and emergency planning, traffic simulation models must be able to capture both the behavior of potential evacuees—if, when, where, and how to evacuate—and the dynamics of the transportation network as congestion forms and dissipates. This chapter described modeling approaches to forecast these decisions, and how simulation tools built on these principles can support advance planning for evacuations.

Future work should continue to refine these behavioral and traffic models, making use of emerging technologies and data sources to better understand and influence evacuation-related travel decisions. The rise of shared mobility services, and the advent of connected and automated vehicles, also present new opportunities for better managing large-scale evacuations. Research in large-scale optimization is also proceeding rapidly, building on advances in artificial intelligence and machine learning. These new data and methodological tools can further enhance the network models described in this chapter, and allow the transportation infrastructure to better serve the unique needs of evacuations.

Practical Implications

Transportation network models are a powerful tool for planning evacuations, serving as a "what if" tool for developing and comparing alternative scenarios and plans. They have significant data and computation requirements, and should be put into place well in advance of a forecasted disaster. Building such a model requires detailed data on the inhabitants of a region: their locations, and information relating to how they will receive and respond to evacuation instructions (sociodemographic factors). Such models also require detailed data on the physical transportation infrastructure, roadway locations, capacities, and susceptibility to disruption (e.g., low water crossings). Data validation is crucial, because even small systematic errors in either data set can lead to significant errors in scenario evaluation and comparison (Boyles & Ruiz Juri, 2019). Specific scenarios can then be built considering a specific region, a set of potential disasters, and alternative management strategies (information dissemination, evacuation timing and staging, and so forth),

Discussion Questions

1. In what ways do travel decisions made during disasters differ from travel decisions made under ordinary circumstances? How can we understand decisions that are only made under rare and exceptional circumstances, and therefore with much less available data?

2. How can traffic simulation and network models be integrated into the emergency planning workflow? Which stakeholders have the data, computing power, and expertise to deploy large-scale simulation models?
3. How can these network models be validated in the field? What kinds of data should be gathered before, during, and after an evacuation in order to gauge the accuracy of a simulation model? How might a simulation model be improved after an event using this data?

Note

1 Loosely speaking, NP-hard problems are those for which efficient solution methods are not known.

References

Abdelgawad, H., & Abdulhai, B. (2009). Emergency evacuation planning as a network design problem: A critical review. *Transportation Letters, 1*, 41–58. https://dx.doi.org/10.3328/TL.2009.01.01.41–58

Abdelgawad, H., Abdulhai, B., & Wahba, M. (2010). Multiobjective optimization for multimodal evacuation. *Transportation Research Record: Journal of the Transportation Research Board, 2196*, 21–33. https://dx.doi.org/10.3141/2196-03

Akbarzadeh, M., & Wilmot, C. G. (2014). Time-dependent route choice in hurricane evacuation. *Natural Hazards Review, 16*. https://dx.doi.org/10.1061/(ASCE)NH.1527-6996.0000159

Alçada-Almeida, L., Tralhão, L., Santos, L., & Coutinho-Rodrigues, J. (2009). A multiobjective approach to locate emergency shelters and identify evacuation routes in urban areas. *Geographical Analysis, 41*, 9–29. https://doi.org/10.1111/j.1538-4632.2009.00745.x

An, S., Cui, N., Li, X., & Ouyang, Y. (2013). Location planning for transit-based evacuation under the risk of service disruptions. *Transportation Research Part B: Methodological, 54*, 1–16. https://dx.doi.org/10.1016/j.trb.2013.03.002

Arlikatti, S., Lindell, M. K., Prater, C. S., & Zhang, Y. (2006). Risk area accuracy and hurricane evacuation expectations of coastal residents. *Environment and Behavior, 38*, 226–247. https://dx.doi.org/10.1177/0013916505277603

Ayfadopoulou, G., Stamos, I., Mitsakis, E., & Grau, J. M. S. (2012). Dynamic traffic assignment based evacuation planning for CBD areas. *Procedia—Social and Behavioral Sciences, 48*, 1078–1087. https://dx.doi.org/10.1016/j.sbspro.2012.06.1084

Baker, E. J. (1991). Hurricane evacuation behavior. *International Journal of Mass Emergencies and Disasters, 9*, 287–310. http://ijmed.org/articles/412/

Barrett, B., Ran, B., & Pillai, R. (2000). Developing a dynamic traffic management modeling framework for hurricane evacuation. *Transportation Research Record: Journal of the Transportation Research Board, 1733*, 115–121. https://dx.doi.org/10.3141/1733-15

Bateman, J. M., & Edwards, B. (2002). Gender and evacuation: A closer look at why women are more likely to evacuate for hurricanes. *Natural Hazards Review, 3*, 107–117. https://dx.doi.org/10.1061/(asce)1527-6988(2002)3:3(107)

Bayram, V., Tansel, B. Ç., & Yaman, H. (2015). Compromising system and user interests in shelter location and evacuation planning. *Transportation Research Part B: Methodological, 72*, 146–163. https://dx.doi.org/10.1016/j.trb.2014.11.010

Bian, R., & Wilmot, C. G. (2018). An analysis on transit pick-up points for vulnerable people during hurricane evacuation: A case study of New Orleans. *International Journal of Disaster Risk Reduction, 31*, 1143–1151. https://dx.doi.org/10.1016/j.ijdrr.2017.07.005

Bish, D. R. (2011). Planning for a bus-based evacuation. *OR Spectrum, 33*, 629–654. https://dx.doi.org/10.1007/s00291-011-0256-1

Bish, D. R., Agca, E., & Glick, R. (2014). Decision support for hospital evacuation and emergency response. *Annals of Operations Research, 221*, 89–106. https://dx.doi.org/10.1007/s10479-011-0943-y

Bish, D. R., & Sherali, H. D. (2013). Aggregate-level demand management in evacuation planning. *European Journal of Operational Research, 224*, 79–92. https://dx.doi.org/10.1016/j.ejor.2012.07.036

Boyles, S. D., & Ruiz Juri, N. (2019). Queue spillback and demand uncertainty in dynamic network loading. *Transportation Research Record.* https://doi.org/10.1177/0361198119826023

Boyles, S. D., & Waller, S. T. (2011a). Traffic network analysis and design. *Wiley Encyclopedia of Operations Research and Management Science.* John Wiley & Sons, Inc. https://dx.doi.org/10.1002/9780470400531.eorms0915

Boyles, S. D., & Waller, S. T. (2011b). Optimal information location for adaptive routing. *Networks and Spatial Economics, 11*, 233–254. https://doi.org/10.1007/s11067-009-9108-9

Brodie, M., Weltzien, E., Altman, D., Blendon, R. J., & Benson, J. M. (2006). Experiences of Hurricane Katrina evacuees in Houston shelters: Implications for future planning. *American Journal of Public Health, 96*, 1402–1408. https://dx.doi.org/10.2105/ajph.2005.084475

Brown, C., White, W., van Slyke, C., & Benson, J. D. (2009). Development of a strategic hurricane evacuation–dynamic traffic assignment model for the Houston, Texas, region. *Transportation Research Record: Journal of the Transportation Research Board, 2137*(1), 46–53. https://dx.doi.org/10.3141/2137-06

Bruneau, M., Chang, S. E., Eguchi, R. T., Lee, G. C., O'Rourke, T. D., Reinhorn, A. M., Shinouzka, M., Tierney, K., Wallace, W. A., von Winterfeldt, D. (2003). A framework to quantitatively assess and enhance the seismic resilience of communities. *Earthquake Spectra, 19*, 733–752. https://dx.doi.org/10.1193/1.1623497

Carnegie, J., & Deka, D. (2010). Using hypothetical disaster scenarios to predict evacuation behavioral response. In *Proceedings of the 87th Annual Meeting Transportation Research Board.* Washington, DC, USA.

Çelik, M., Ergun, Ö., Johnson, B., Keskinocak, P., Lorca, Á., Pekgün, P., & Swann, J. (2012). Humanitarian logistics. In *2012 TutORials in Operations Research* (pp. 18–49). INFORMS. https://dx.doi.org/10.1287/educ.1120.0100

Çelik, M., Ergun, Ö., & Keskinocak, P. (2015). The post-disaster debris clearance problem under incomplete information. *Operations Research, 63*, 65–85. https://dx.doi.org/10.1287/opre.2014.1342

Chang, S. E., & Nojima, N. (2001). Measuring post-disaster transportation system performance: The 1995 Kobe earthquake in comparative perspective. *Transportation Research Part A: Policy and Practice, 35*, 475–494. https://dx.doi.org/10.1016/s0965-8564(00)00003-3

Chen, C.-C., & Chou, C.-S. (2009). Modeling and performance assessment of a transit-based evacuation plan within a contraflow simulation environment. *Transportation Research Record: Journal of the Transportation Research Board, 2091*, 40–50. https://dx.doi.org/10.3141/2091-05

Chen, L., & Miller-Hooks, E. (2012). Resilience: An indicator of recovery capability in intermodal freight transport. *Transportation Science, 46,* 109–123. https://dx.doi.org/10.1287/trsc.1110.0376

Chen, M., Chen, L., & Miller-Hooks, E. (2007). Traffic signal timing for urban evacuation. *Journal of Urban Planning and Development, 133,* 30–42. https://dx.doi.org/10.1061/(asce)0733-9488(2007)133:1(30)

Chen, X., & Zhan, F. B. (2008). Agent-based modelling and simulation of urban evacuation: Relative effectiveness of simultaneous and staged evacuation strategies. *Journal of the Operational Research Society, 59,* 25–33. https://dx.doi.org/10.1057/palgrave.jors.2602321

Cheng, G. (2007). Friction factor function calibration for hurricane evacuation trip distribution. In *Proceedings of the 87th Annual Meeting Transportation Research Board.* Washington, DC.

Cheng, G., Wilmot, C. G., & Baker, E. J. (2008). A destination choice model for hurricane evacuation. In *Proceedings of the 87th Annual Meeting Transportation Research Board.* Washington, DC.

Cheng, G., Wilmot, C. G., & Baker, E. J. (2011). Dynamic gravity model for hurricane evacuation planning. *Transportation Research Record: Journal of the Transportation Research Board, 2234,* 125–134. https://dx.doi.org/10.3141/2234-14

Chiu, Y.-C., Bottom, J., Mahut, M., Paz, A., Balakrishna, R., Waller, T., & Hicks, J. (2011). Dynamic traffic assignment: A primer. *Transportation Research E-Circular* (E-C153). Transportation Research Board.

Chiu, Y.-C., & Mirchandani, P. B. (2008). Online behavior-robust feedback information routing strategy for mass evacuation. *IEEE Transactions on Intelligent Transportation Systems, 9,* 264–274. https://dx.doi.org/10.1109/tits.2008.922878

Coutinho-Rodrigues, J., Tralhão, L., & Alçada-Almeida, L. (2012). Solving a location-routing problem with a multiobjective approach: The design of urban evacuation plans. *Journal of Transport Geography, 22,* 206–218. https://dx.doi.org/10.1016/j.jtrangeo.2012.01.006

Cova, T. J., Dennison, P. E., & Drews, F. A. (2011). Modeling evacuate versus shelter-in-place decisions in Wildfires. *Sustainability, 3,* 1662–1687. https://dx.doi.org/10.3390/su3101662

Cova, T. J., & Johnson, J. P. (2003). A network flow model for lane-based evacuation routing. *Transportation Research Part A: Policy and Practice, 37,* 579–604. https://dx.doi.org/10.1016/s0965-8564(03)00007-7

Cuellar, L., Kubicek, D., Hengartner, N., & Hansson, A. (2009). Emergency relocation: Population response model to disasters. In *2009 IEEE Conference on Technologies for Homeland Security.* IEEE. https://dx.doi.org/10.1109/ths.2009.5168096

Deka, D., & Carnegie, J. (2010). *Analyzing evacuation behavior of transportation-disadvantaged populations in northern New Jersey.* Presented at the 89th Annual Meeting of the Transportation Research Board, Washington, DC.

Dennison, P. E., Cova, T. J., & Mortiz, M. A. (2006). WUIVAC: A wildland-urban interface evacuation trigger model applied in strategic wildfire scenarios. *Natural Hazards, 41,* 181–199. https://dx.doi.org/10.1007/s11069-006-9032-y

Dixit, V., Montz, T., & Wolshon, B. (2011). Validation techniques for region-level microscopic mass evacuation traffic simulations. *Transportation Research Record: Journal of the Transportation Research Board, 2229*(1), 66–74. https://dx.doi.org/10.3141/2229-08

Dixit, V. V., & Radwan, E. (2009). Hurricane evacuation: Origin, route, and destination. *Journal of Transportation Safety & Security, 1,* 74–84. https://dx.doi.org/10.1080/19439960902735048

Donovan, B., & Work, D. B. (2017). Empirically quantifying city-scale transportation system resilience to extreme events. *Transportation Research Part C: Emerging Technologies, 79,* 333–346. https://dx.doi.org/10.1016/j.trc.2017.03.002

Dow, K., & Cutter, S. L. (1998). Crying wolf: Repeat responses to hurricane evacuation orders. *Coastal Management, 26,* 237–252. https://dx.doi.org/10.1080/08920759809362356

Dow, K., & Cutter, S. L. (2000). Public orders and personal opinions: Household strategies for hurricane risk assessment. *Global Environmental Change Part B: Environmental Hazards, 2,* 143–155. https://dx.doi.org/10.1016/s1464-2867(01)00014-6

Dow, K., & Cutter, S. L. (2002). Emerging hurricane evacuation issues: Hurricane Floyd and South Carolina. *Natural Hazards Review, 3,* 12–18. https://dx.doi.org/10.1061/(asce)1527-6988(2002)3:1(12)

Duran, S., Ergun, Ö., Keskinocak, P., & Swann, J. L. (2012). Humanitarian logistics: Advanced purchasing and pre-positioning of relief items. In J. Bookbinder (Ed.), *Handbook of Global Logistics* (pp. 447–462). New York, NY: Springer.

Elliott, J. R., & Pais, J. (2006). Race, class, and Hurricane Katrina: Social differences in human responses to disaster. *Social Science Research, 35,* 295–321. https://dx.doi.org/10.1016/j.ssresearch.2006.02.003

FEMA (1999). *Northwest Florida Hurricane evacuation study: Technical data report for Escambia, Santa Rosa, Okaloosa, Walton, Bay, Holmes, Jackson, and Washington Counties, Florida.* Florida State Emergency Management Office. Retrieved from: https://coast.noaa.gov/

Fu, H., & Wilmot, C. (2004). Sequential logit dynamic travel demand model for hurricane evacuation. *Transportation Research Record: Journal of the Transportation Research Board, 1882,* 19–26. https://dx.doi.org/10.3141/1882-03

Fu, H., Wilmot, C. G., Zhang, H., & Baker, E. J. (2007). Modeling the hurricane evacuation response curve. *Transportation Research Record: Journal of the Transportation Research Board, 2022,* 94–102. https://dx.doi.org/10.3141/2022-11

Gao, S. (2012). Modeling strategic route choice and real-time information impacts in stochastic and time-dependent networks. *IEEE Transactions on Intelligent Transportation Systems, 13,* 1298–1311. https://dx.doi.org/10.1109/tits.2012.2187197

Gibbs, L., & Holloway, C. (2013). *Hurricane Sandy after action: Report and recommendations to Mayor Michael R. Bloomberg.* Retrieved from: www1.nyc.gov/assets/housingrecovery/downloads/pdf/2017/sandy_aar_5-2-13.pdf

Golshani, N., Shabanpour, R., Auld, J., Mohammadian, A. K., & Ley, H. (2018). *Modeling evacuation destination and departure time choices for no-notice emergency events.* Presented at the 97th Annual Meeting of the Transportation Research Board, Washington, DC.

Gray, B. H., & Hebert, K. (2007). Hospitals in Hurricane Katrina: Challenges facing custodial institutions in a disaster. *Journal of Health Care for the Poor and Underserved, 18,* 283–298. https://dx.doi.org/10.1353/hpu.2007.0031

Gudishala, R., & Wilmot, C. (2012). Comparison of time-dependent sequential logit and nested logit for modeling hurricane evacuation demand. *Transportation Research Record: Journal of the Transportation Research Board, 2312,* 134–140. https://dx.doi.org/10.3141/2312-14

Hadas, Y., & Laor, A. (2013). Network design model with evacuation constraints. *Transportation Research Part A: Policy and Practice, 47,* 1–9. https://dx.doi.org/10.1016/j.tra.2012.10.027

Hamza-Lup, G. L., Hua, K. A., Le, M., & Peng, R. (2008). Dynamic plan generation and real time management techniques for traffic evacuation. *IEEE Transactions on Intelligent Transportation Systems, 9*, 615–624. https://dx.doi.org/10.1109/tits.2008.2006738

Han, L. D., Yuan, F., Chin, S.-M., & Hwang, H. (2006). Global optimization of emergency evacuation assignments. *Interfaces, 36*, 502–513. https://dx.doi.org/10.1287/inte.1060.0251

Hasan, S., Mesa-Arango, R., Ukkusuri, S., & Murray-Tuite, P. (2012). Transferability of hurricane evacuation choice model: Joint model estimation combining multiple data sources. *Journal of Transportation Engineering, 138*, 548–556. https://dx.doi.org/10.1061/(asce)te.1943-5436.0000365

Hasan, S., & Ukkusuri, S. V. (2011). A threshold model of social contagion process for evacuation decision making. *Transportation Research Part B: Methodological, 45*, 1590–1605. https://dx.doi.org/10.1016/j.trb.2011.07.008

Hasan, S., Ukkusuri, S., Gladwin, H., & Murray-Tuite, P. (2011). Behavioral model to understand household-level hurricane evacuation decision making. *Journal of Transportation Engineering, 137*, 341–348. https://dx.doi.org/10.1061/(asce)te.1943-5436.0000223

Hausknecht, M., Au, T.-C., Stone, P., Fajardo, D., & Waller, T. (2011). Dynamic lane reversal in traffic management. In *2011 14th International IEEE Conference on Intelligent Transportation Systems (ITSC)*. IEEE. https://dx.doi.org/10.1109/itsc.2011.6082932

He, S., L. Zhang, R., Song, Y. Wen, & Wu, D. (2009). Optimal transit routing problem for emergency evacuations. In *Proceedings of the 88th Annual Meeting Transportation Research Board*. Washington, DC.

Hobeika, A. G., & Kim, C. (1998). Comparison of traffic assignments in evacuation modeling. *IEEE Transactions on Engineering Management, 45*, 192–198. https://dx.doi.org/10.1109/17.669768

Holguín-Veras, J., & Jaller, M. (2011). Immediate resource requirements after Hurricane Katrina. *Natural Hazards Review, 13*, 117–131. https://dx.doi.org/10.1061/(asce)nh.1527-6996.0000068

Holguín-Veras, J., Taniguchi, E., Jaller, M., Aros-Vera, F., Ferreira, F., & Thompson, R. G. (2014). The Tohoku disasters: Chief lessons concerning the post disaster humanitarian logistics response and policy implications. *Transportation Research Part A: Policy and Practice, 69*, 86–104. https://dx.doi.org/10.1016/j.tra.2014.08.003

Hsu, Y.-T., & Peeta, S. (2014). Risk-based spatial zone determination problem for stage-based evacuation operations. *Transportation Research Part C: Emerging Technologies, 41*, 73–89. https://dx.doi.org/10.1016/j.trc.2014.01.013

Hua, J., Dai, L., Wang, Y., & Li, Y. (2016). Evacuation network optimization with a transit staged strategy. In *CICTP 2016, American Society of Civil Engineers*. https://dx.doi.org/10.1061/9780784479896.155

Huang, S.-K., Lindell, M. K., & Prater, C. S. (2016). Who leaves and who stays? A review and statistical meta-analysis of hurricane evacuation studies. *Environment and Behavior, 48*, 991–1029. https://dx.doi.org/10.1177/0013916515578485

Jafari, E., Pandey, V., & Boyles, S. D. (2017). A decomposition approach to the static traffic assignment problem. *Transportation Research Part B, 105*, 270–296. https://doi.org/10.1016/j.trb.2017.09.011

Jelesnianski, C. P., Chen, J., & Shaffer, W. A. (1992). *Sea, lake, and overland surges from hurricanes (SLOSH)*. Retrieved from: www.nhc.noaa.gov/surge/slosh.php

Kalafatas, G., & Peeta, S. (2009). Planning for evacuation: Insights from an efficient network design model. *Journal of Infrastructure Systems, 15*, 21–30. https://dx.doi.org/10.1061/(asce)1076-0342(2009)15:1(21)

Kaufman, S., Qing, C., Levenson, N., & Hanson, M. (2012). *Transportation during and after Hurricane Sandy.* Retrieved from: https://wagner.nyu.edu/files/faculty/publications/sandytransportation.pdf

Kılcı, F., Kara, B. Y., & Bozkaya, B. (2015). Locating temporary shelter areas after an earthquake: A case for Turkey. *European Journal of Operational Research, 243*, 323–332. https://dx.doi.org/10.1016/j.ejor.2014.11.035

Kim, S., Shekhar, S., & Min, M. (2008). Contraflow transportation network reconfiguration for evacuation route planning. *IEEE Transactions on Knowledge and Data Engineering, 20*, 1115–1129. https://dx.doi.org/10.1109/tkde.2007.190722

Kirschenbaum, A. (1992). Warning and evacuation during a mass disaster: A multivariate decision-making model. *International Journal of Mass Emergencies and Disasters, 10*, 91–114.

Knoop, V. L., Hoogendoorn, S. P., & van Zuylen, H. (2010). Rerouting behaviour of travellers under exceptional traffic conditions—an empirical analysis of route choice. *Procedia Engineering, 3*, 113–128. https://dx.doi.org/10.1016/j.proeng.2010.07.012

Kongsomsaksakul, S., Yang, C., & Chen, A. (2005). Shelter location-allocation model for flood evacuation planning. *Journal of the Eastern Asia Society for Transportation Studies, 6*, 4237–4252. https://dx.doi.org/10.11175/easts.6.4237

Kovács, G., & Spens, K. M. (2007). Humanitarian logistics in disaster relief operations. *International Journal of Physical Distribution & Logistics Management, 37*, 99–114. https://dx.doi.org/10.1108/09600030710734820

Kulshrestha, A., Lou, Y., & Yin, Y. (2012). Pick-up locations and bus allocation for transit-based evacuation planning with demand uncertainty. *Journal of Advanced Transportation, 48*, 721–733. https://dx.doi.org/10.1002/atr.1221

Kulshrestha, A., Wu, D., Lou, Y., & Yin, Y. (2011). Robust shelter locations for evacuation planning with demand uncertainty. *Journal of Transportation Safety & Security, 3*, 272–288. https://dx.doi.org/10.1080/19439962.2011.609323

Landsea, C. (2015, June 1). Frequently asked questions. *Atlantic Oceanographic & Meteorological Laboratory.* Retrieved from: www.aoml.noaa.gov/hrd/tcfaq/E11.html

Leach, J. (1994). *Survival psychology.* London: Macmillan.

Levin, M. W., & Boyles, S. D. (2015). Effects of autonomous vehicle ownership on trip, mode, and route choice. *Transportation Research Record, 2493*, 29–38. https://experts.umn.edu/en/publications/effects-of-autonomous-vehicle-ownership-on-trip-mode-and-route-ch

Li, A. C. Y., Nozick, L., Xu, N., & Davidson, R. (2012). Shelter location and transportation planning under hurricane conditions. *Transportation Research Part E: Logistics and Transportation Review, 48*, 715–729. https://dx.doi.org/10.1016/j.tre.2011.12.004

Li, A. C. Y., Xu, N., Nozick, L., & Davidson, R. (2011). Bilevel optimization for integrated shelter location analysis and transportation planning for hurricane events. *Journal of Infrastructure Systems, 17*, 184–192. https://dx.doi.org/10.1061/(asce)is.1943-555x.0000067

Li, L., Jin, M., & Zhang, L. (2011). Sheltering network planning and management with a case in the Gulf Coast region. *International Journal of Production Economics, 131*, 431–440. https://dx.doi.org/10.1016/j.ijpe.2010.12.013

Li, Q., Yang, X. K., & Wei, H. (2006). Integrating traffic simulation models with evacuation planning system in a GIS environment. In *2006 IEEE Intelligent Transportation Systems Conference*. IEEE. https://dx.doi.org/10.1109/itsc.2006.1706805

Liberatore, F., Ortuño, M. T., Tirado, G., Vitoriano, B., & Scaparra, M. P. (2014). A hierarchical compromise model for the joint optimization of recovery operations and distribution of emergency goods in humanitarian logistics. *Computers & Operations Research, 42*, 3–13. https://dx.doi.org/10.1016/j.cor.2012.03.019

Lim, E., & Wolshon, B. (2005). Modeling and performance assessment of contraflow evacuation termination points. *Transportation Research Record: Journal of the Transportation Research Board, 1922*, 118–128. https://dx.doi.org/10.3141/1922-16

Lin, D.-Y., Eluru, N., Waller, S. T., & Bhat, C. R. (2009). Evacuation planning using the integrated system of activity-based modeling and dynamic traffic assignment. *Transportation Research Record: Journal of the Transportation Research Board, 2132*, 69–77. https://dx.doi.org/10.3141/2132-08

Lindell, M. K. (2008). EMBLEM2: An empirically based large scale evacuation time estimate model. *Transportation Research Part A: Policy and Practice, 42*, 140–154. https://dx.doi.org/10.1016/j.tra.2007.06.014

Lindell, M. K., Kang, J. E., & Prater, C. S. (2011). The logistics of household hurricane evacuation. *Natural Hazards, 58*, 1093–1109. https://dx.doi.org/10.1007/s11069-011-9715-x

Lindell, M. K., Lu, J.-C., & Prater, C. S. (2005). Household decision making and evacuation in response to Hurricane Lili. *Natural Hazards Review, 6*, 171–179. https://dx.doi.org/10.1061/(asce)1527-6988(2005)6:4(171)

Lindell, M. K. and Perry, R. W. (2003). *Communicating environmental risk in multiethnic Communities* (Vol. 7). Thousand Oaks, CA: Sage.

Lindell, M. K., & Prater, C. S. (2007). Critical behavioral assumptions in evacuation time estimate analysis for private vehicles: Examples from hurricane research and planning. *Journal of Urban Planning and Development, 133*, 18–29. https://dx.doi.org/10.1061/(asce)0733-9488(2007)133:1(18)

Lindell, M. K., Prater, C. S., Wu, H. C., Huang, S.-K., Johnston, D. M., Becker, J. S., & Shiroshita, H. (2015). Immediate behavioural responses to earthquakes in Christchurch, New Zealand, and Hitachi, Japan. *Disasters, 40*, 85–111. https://dx.doi.org/10.1111/disa.12133

Litman, T. (2006). Lessons from Katrina and Rita: What major disasters can teach transportation planners. *Journal of Transportation Engineering, 132*, 11–18. https://dx.doi.org/10.1061/(asce)0733-947x(2006)132:1(11)

Liu, S., Murray-Tuite, P., & Schweitzer, L. (2011). Relocating children in daytime no-notice evacuations: Methodology and applications for transport systems of personal vehicles and buses. *Transportation Research Record: Journal of the Transportation Research Board, 2234*, 79–88. https://dx.doi.org/10.3141/2234-09

Mesa-Arango, R., Hasan, S., Ukkusuri, S. V., & Murray-Tuite, P. (2013). Household-level model for hurricane evacuation destination type choice using Hurricane Ivan data. *Natural Hazards Review, 14*, 11–20. https://dx.doi.org/10.1061/(asce)nh.1527-6996.0000083

Mileti, D. S., & O'Brien, P. W. (1992). Warnings during disaster: Normalizing communicated risk. *Social Problems, 39*, 40–57. https://dx.doi.org/10.2307/3096912

Mileti, D. S., & Peek, L. (2000). The social psychology of public response to warnings of a nuclear power plant accident. *Journal of Hazardous Materials, 75*, 181–194. https://dx.doi.org/10.1016/s0304-3894(00)00179-5

Mileti, D. S., & Sorensen, J. H. (1987). Natural hazards and precautionary behavior. In N. Weinstein (Ed.), *Taking care: Understanding and encouraging self-protective behavior* (pp. 189–207). Cambridge: Cambridge University Press.

Mudigonda, S., Ozbay, K., & Bartin, B. (2018). Evaluating the resilience and recovery of public transit system using big data: Case study from New Jersey. *Journal of Transportation Safety & Security.* https://dx.doi.org/10.1080/19439962.2018.1436105

Murray-Tuite, P., & Mahmassani, H. (2003). Model of household trip-chain sequencing in emergency evacuation. *Transportation Research Record: Journal of the Transportation Research Board, 1831,* 21–29. https://dx.doi.org/10.3141/1831-03

Murray-Tuite, P., & Mahmassani, H. (2004). Transportation network evacuation planning with household activity interactions. *Transportation Research Record: Journal of the Transportation Research Board, 1894,* 150–159. https://dx.doi.org/10.3141/1894-16

Murray-Tuite, P., & Wolshon, B. (2013). Evacuation transportation modeling: An overview of research, development, and practice. *Transportation Research Part C: Emerging Technologies, 27,* 25–45. https://doi.org/10.1016/j.trc.2012.11.005

Murray-Tuite, P., Yin, W., Ukkusuri, S. V., & Gladwin, H. (2012). Changes in evacuation decisions between Hurricanes Ivan and Katrina. *Transportation Research Record: Journal of the Transportation Research Board, 2312,* 98–107. https://dx.doi.org/10.3141/2312-10

Naghawi, H., & Wolshon, B. (2010). Transit-based emergency evacuation simulation modeling. *Journal of Transportation Safety & Security, 2,* 184–201. https://dx.doi.org/10.1080/19439962.2010.488316

Naghawi, H., & Wolshon, B. (2012). Performance of traffic networks during multimodal evacuations: Simulation-based assessment. *Natural Hazards Review, 13,* 196–204. https://dx.doi.org/10.1061/(asce)nh.1527-6996.0000065

Nahum, O. E., & Hadas, Y. (2017). Multi-objective evacuation network design with chance constraints. In *Transportation Research Board (TRB) 96th Annual Meeting.* Washington DC.

Ng, M., Park, J., & Waller, S. T. (2010). A hybrid bilevel model for the optimal shelter assignment in emergency evacuations. *Computer-Aided Civil and Infrastructure Engineering, 25,* 547–556. https://dx.doi.org/10.1111/j.1467-8667.2010.00669.x

Overstreet, R. E., Hall, D., Hanna, J. B., & Rainer, R. K., Jr. (2011). Research in humanitarian logistics. *Journal of Humanitarian Logistics and Supply Chain Management, 1,* 114–131. https://dx.doi.org/10.1108/20426741111158421

Peeta, S., & Ziliaskopoulos, A. K. (2001). Foundations of dynamic traffic assignment: The past, the present and the future. *Networks and Spatial Economics, 1,* 233–265. https://dx.doi.org/10.1023/a:1012827724856

Pel, A. J., Bliemer, M. C. J., & Hoogendoorn, S. P. (2008). EVAQ: A new analytical model for voluntary and mandatory evacuation strategies on time-varying networks. In *2008 11th International IEEE Conference on Intelligent Transportation Systems.* IEEE. https://dx.doi.org/10.1109/itsc.2008.4732655

Pel, A. J., Bliemer, M. C. J., & Hoogendoorn, S. P. (2009). Hybrid route choice modeling in dynamic traffic assignment. *Transportation Research Record: Journal of the Transportation Research Board, 2091,* 100–107. https://dx.doi.org/10.3141/2091-11

Pel, A. J., Bliemer, M. C. J., & Hoogendoorn, S. P. (2011). A review on travel behaviour modelling in dynamic traffic simulation models for evacuations. *Transportation, 39,* 97–123. https://dx.doi.org/10.1007/s11116-011-9320-6

Pel, A. J., Hoogendoorn, S. P., & Bliemer, M. C. J. (2010). Evacuation modeling including traveler information and compliance behavior. *Procedia Engineering, 3,* 101–111. https://dx.doi.org/10.1016/j.proeng.2010.07.011

Pidd, M., de Silva, F. N., & Eglese, R. W. (1993). Cemps: Configurable evacuation management and planning system—a progress report. In *Proceedings of 1993 Winter Simulation Conference (WSC '93)*. IEEE. https://dx.doi.org/10.1109/wsc.1993.718397

Pidd, M., de Silva, F. N., & Eglese, R. W. (1996). A simulation model for emergency evacuation. *European Journal of Operational Research, 90*, 413–419. https://dx.doi.org/10.1016/0377-2217(95)00112-3

Rambha, T., Boyles, S. D., Unnikrishnan, A., & Stone, P. (2018). Marginal cost pricing for system optimal traffic assignment with recourse under supply-side uncertainty. *Transportation Research Part B: Methodological, 110*, 104–121. https://dx.doi.org/10.1016/j.trb.2018.02.008

Rambha, T., Boyles, S. D., & Waller, S. T. (2016) Adaptive transit routing in stochastic time-dependent networks. *Transportation Science, 50*, 1043–1059. https://doi.org/10.1287/trsc.2015.0613

Rathi, A. K., & Solanki, R. S. (1993). Simulation of traffic flow during emergency evacuations: a microcomputer based modeling system. In *Proceedings of 1993 Winter Simulation Conference (WSC '93)*. IEEE. https://dx.doi.org/10.1109/wsc.1993.718387

Rebennack, S., Arulselvan, A., Elefteriadou, L., & Pardalos, P. M. (2008). Complexity analysis for maximum flow problems with arc reversals. *Journal of Combinatorial Optimization, 19*, 200–216. https://dx.doi.org/10.1007/s10878-008-9175-8

Robinson, R. M., & Khattak, A. (2010). Route change decision making by hurricane evacuees facing congestion. *Transportation Research Record: Journal of the Transportation Research Board, 2196*, 168–175. https://dx.doi.org/10.3141/2196-18

Rui, S., Shiwei, H., & Zhang, L. (2009). Optimum transit operations during the emergency evacuations. *Journal of Transportation Systems Engineering and Information Technology, 9*, 154–160. https://dx.doi.org/10.1016/S1570-6672(08)60096-3

Sadri, A. M., Ukkusuri, S. V., Murray-Tuite, P., & Gladwin, H. (2014). How to evacuate: Model for understanding the routing strategies during hurricane evacuation. *Journal of Transportation Engineering, 140*, 61–69. https://dx.doi.org/10.1061/(asce)te.1943-5436.0000613

Sarwar, M. T., Anastasopoulos, P. C., Ukkusuri, S. V., Murray-Tuite, P., & Mannering, F. L. (2016). A statistical analysis of the dynamics of household hurricane-evacuation decisions. *Transportation, 45*, 51–70. https://dx.doi.org/10.1007/s11116-016-9722-6

Sayyady, F., & Eksioglu, S. D. (2010). Optimizing the use of public transit system during no-notice evacuation of urban areas. *Computers & Industrial Engineering, 59*, 488–495. https://dx.doi.org/10.1016/j.cie.2010.06.001

Serulle, N. U., & Cirillo, C. (2017). The optimal time to evacuate: A behavioral dynamic model on Louisiana resident data. *Transportation Research Part B: Methodological, 106*, 447–463. https://dx.doi.org/10.1016/j.trb.2017.06.004

Shahparvari, S., & Abbasi, B. (2017). Robust stochastic vehicle routing and scheduling for bushfire emergency evacuation: An Australian case study. *Transportation Research Part A: Policy and Practice, 104*, 32–49. https://dx.doi.org/10.1016/j.tra.2017.04.036

Shahparvari, S., Chhetri, P., Abbasi, B., & Abareshi, A. (2016). Enhancing emergency evacuation response of late evacuees: Revisiting the case of Australian Black Saturday bushfire. *Transportation Research Part E: Logistics and Transportation Review, 93*, 148–176. https://dx.doi.org/10.1016/j.tre.2016.05.010

Sheffi, Y., Mahmassani, H., & Powell, W. B. (1982). A transportation network evacuation model. *Transportation Research Part A: General, 16*, 209–218. https://dx.doi.org/10.1016/0191-2607(82)90022-x

Sherali, H. D., Carter, T. B., & Hobeika, A. G. (1991). A location-allocation model and algorithm for evacuation planning under hurricane/flood conditions. *Transportation Research Part B: Methodological, 25,* 439–452. https://dx.doi.org/10.1016/0191-2615(91)90037-j

Sivak, M. (2013). *Has motorization in the US peaked?* Technical report, University of Michigan Transportation Research Institute.

Smith, S. K., & McCarty, C. (2009). Fleeing the storm(s): An examination of evacuation behavior during Florida's 2004 hurricane season. *Demography, 46*(1), 127–145.

Soika, K. (2006). Evacuation planning in Texas before and after Hurricane Rita. *House Research Organization, Texas House of Representatives, Interim News, 79*(2), 1–8.

Sorensen, J. H. (2000). Hazard warning systems: Review of 20 years of progress. *Natural Hazards Review, 1,* 119–125. https://dx.doi.org/10.1061/(asce)1527-6988(2000)1:2(119)

Stephens, K. K., Jafari, E., Boyles, S., Ford, J. L., & Zhu, Y. (2015). Increasing evacuation communication through ICTs: An agent-based model demonstrating evacuation practices and the resulting traffic congestion in the rush to the road. *Journal of Homeland Security and Emergency Management, 12,* 497–528. https://dx.doi.org/10.1515/jhsem-2014-0075

Sutton, J. N., Palen, L., & Shklovski, I. (2008). Backchannels on the front lines: Emergency uses of social media in the 2007 Southern California Wildfires. In F. Fiedrich & B. Van de Walle (Eds.), *Proceedings of the 5th International ISCRAM Conference.*

Swamy, R., Kang, J. E., Batta, R., & Chung, Y. (2017). Hurricane evacuation planning using public transportation. *Socio-Economic Planning Sciences, 59,* 43–55. https://doi.org/10.1016/j.seps.2016.10.009

Theodoulou, G., & Wolshon, B. (2004). Alternative methods to increase the effectiveness of freeway contraflow evacuation. *Transportation Research Record: Journal of the Transportation Research Board, 1865,* 48–56. https://doi.org/10.3141/1865-08

Toth, P., & Vigo, D. (2002). *The vehicle routing problem.* SIAM. https://doi.org/10.1137/1.9781611973594

Troncoso Parady, G., & Hato, E. (2016). Accounting for spatial correlation in tsunami evacuation destination choice: a case study of the Great East Japan Earthquake. *Natural Hazards, 84,* 797–807. https://dx.doi.org/10.1007/s11069-016-2457-z

Tuydes, H., & Ziliaskopoulos, A. (2004). *Network re-design to optimize evacuation contraflow.* Presented at the 83rd Annual Meeting of the Transportation Research Board, Washington, DC.

Tuydes, H., & Ziliaskopoulos, A. (2006). Tabu-based heuristic approach for optimization of network evacuation contraflow. *Transportation Research Record: Journal of the Transportation Research Board, 1964,* 157–168. https://doi.org/10.1177/0361198106196400117

Tweedie, S. W., Rowland, J. R., Walsh, S. J., Rhoten, R. P., & Hagle, P. I. (1986). A methodology for estimating emergency evacuation times. *The Social Science Journal, 23,* 189–204. https://doi.org/10.1016/0362-3319(86)90035-2

Ukkusuri, S. V., Hasan, B., Luong, B., Doan, K., Zhan, X., Murray-Tuite, P., & Yin, W. (2017). A-RESCUE: An agent based regional evacuation simulator coupled with user enriched behavior. *Networks and Spatial Economics, 17,* 197–223. https://doi.org/10.1007/s11067-016-9323-0

Unnikrishnan, A., & Waller, S. T. (2009). User equilibrium with recourse. *Networks and Spatial Economics, 9*(4), 575–593. https://doi.org/10.1007/s11067-009-9114-y

Urbina, E., & Wolshon, B. (2003). National review of hurricane evacuation plans and policies: a comparison and contrast of state practices. *Transportation Research Part A: Policy and Practice, 37,* 257–275. https://doi.org/10.1016/s0965-8564(02)00015-0

Villalobos, J., Chiu, Y.-C., Zheng, H., & Gautam, B. (2006). *Modeling and solving optimal evacuation destination-route-flow-staging problem for no-notice extreme events.* Presented at the 85th Annual Meeting of the Transportation Research Board, Washington, DC.

Wardrop, J. G. (1952). Some theoretical aspects of road traffic research. *Proceedings of the Institution of Civil Engineers, 1*, 325–362. https://doi.org/10.1680/ipeds.1952.11259

Wei, H.-L., Wu, H.-C., Lindell, M. K., Prater, C. S., Shiroshita, H., Johnston, S. D., & Becker, J. S. (2017). Assessment of households' responses to the tsunami threat: A comparative study of Japan and New Zealand. *International Journal of Disaster Risk Reduction, 25*, 274–282. https://doi.org/10.1016/j.ijdrr.2017.09.011

Whitehead, J. C., Edwards, B., Van Willigen, M., Maiolo, J. R., Wilson, K., & Smith, K. T. (2000). Heading for higher ground: factors affecting real and hypothetical hurricane evacuation behavior. *Global Environmental Change Part B: Environmental Hazards, 2*, 133–142. https://doi.org/10.3763/ehaz.2000.0219

Wilmot, C. G., & Meduri, N. (2005). Methodology to establish hurricane evacuation zones. *Transportation Research Record: Journal of the Transportation Research Board, 1922*, 129–137. https://doi.org/10.1177/0361198105192200117

Wilmot, C. G., & Mei, B. (2004). Comparison of alternative trip generation models for hurricane evacuation. *Natural Hazards Review, 5*(4), 170–178. https://doi.org/10.1061/(asce)1527-6988(2004)5:4(170)

Wolshon, B. (2001). "One-way-out": Contraflow freeway operation for hurricane evacuation. *Natural Hazards Review, 2*, 105–112. https://doi.org/10.1061/(asce)1527-6988(2001)2:3(105)

Wolshon, B., Lefate, J., Naghawi, H., Montz, T., & Dixit, V. (2009). Application of TRANSIMS for the multimodal microscale simulation of the New Orleans emergency evacuation plan. *Technical report 09-01*, Gulf Coast Center for Evacuation and Transportation Resiliency.

Wolshon, B., & Marchive III, E. (2007). Emergency planning in the urban-wildland interface: Subdivision-level analysis of wildfire evacuations. *Journal of Urban Planning and Development, 133*, 73–81. https://doi.org/10.1061/(asce)0733-9488(2007)133:1(73)

Wolshon, B., Urbina Hamilton, E., Levitan, M., & Wilmot, C. (2005). Review of policies and practices for hurricane evacuation. II: Traffic operations, management, and control. *Natural Hazards Review, 6*, 143–161. https://doi.org/10.1061/(asce)1527-6988(2005)6:3(143)

Wolshon, B., Zhang, Z., Parr, S., Mitchell, B., & Pardue, J. (2015). Agent-based modeling for evacuation traffic analysis in megaregion road networks. *Procedia Computer Science, 52*, 908–913. https://doi.org/10.1016/j.procs.2015.05.164

Wu, H.-C., Lindell, M. K., & Prater, C. S. (2012). Logistics of hurricane evacuation in Hurricanes Katrina and Rita. *Transportation Research Part F: Traffic Psychology and Behavior, 15*, 445–461. https://doi.org/10.1016/j.trf.2012.03.005

Xie, C., Lin, D.-Y., & Waller, S. T. (2010). A dynamic evacuation network optimization problem with lane reversal and crossing elimination strategies. *Transportation Research Part E: Logistics and Transportation Review, 46*(3), 295–316. https://doi.org/10.1016/j.tre.2009.11.004

Xie, C., & Turnquist, M. A. (2009). Integrated evacuation network optimization and emergency vehicle assignment. *Transportation Research Record: Journal of the Transportation Research Board, 2091*, 79–90. https://doi.org/10.3141/2091-09

Xu, K., Davidson, R. A., Nozick, L. K., Wachtendorf, T., & DeYoung, S. E. (2016). Hurricane evacuation demand models with a focus on use for prediction in future events. *Transportation Research Part A: Policy and Practice*, *87*, 90–101. https://doi.org/10.1016/j.tra.2016.02.012

Yahia, C. N., Pandey, V., & Boyles, S. D. (2018) Network partitioning algorithms for solving the traffic assignment problem using a decomposition approach. *Transportation Research Record*. https://doi.org/10.1177/0361198118799039

Yang, H., Morgul, E. F., Ozbay, K., & Xie, K. (2016). Modeling evacuation behavior under hurricane conditions. *Transportation Research Record: Journal of the Transportation Research Board*, *2599*, 63–69. https://doi.org/10.3141/2599-08

Yazici, M. A., & Ozbay, K. (2007). Impact of probabilistic road capacity constraints on the spatial distribution of hurricane evacuation shelter capacities. *Transportation Research Record: Journal of the Transportation Research Board*, *2022*, 55–62. https://doi.org/10.3141/2022-07

Yazici, M. A., & Ozbay, K. (2008). Evacuation modelling in the United States: Does the demand model choice matter? *Transport Reviews*, *28*, 757–779. https://doi.org/10.1080/01441640802041812

Yi, W., Nozick, L., Davidson, R., Blanton, B., & Colle, B. (2017). Optimization of the issuance of evacuation orders under evolving hurricane conditions. *Transportation Research Part B: Methodological*, *95*, 285–304. https://doi.org/10.1016/j.trb.2016.10.008

Yu, J., Pande, A., Nezamuddin, N., Dixit, V., & Edwards, F. (2014). Routing strategies for emergency management decision support systems during evacuation. *Journal of Transportation Safety & Security*, *6*, 257–273. https://doi.org/10.1080/19439962.2013.863258

Yuan, F., Han, L., Chin, S.-M., & Hwang, H. (2006). Proposed framework for simultaneous optimization of evacuation traffic destination and route assignment. *Transportation Research Record: Journal of the Transportation Research Board*, *1964*, 50–58. https://doi.org/10.1177/0361198106196400107

Zeigler, D. J., Brunn, S. D., & Johnson Jr., J. H. (1981). Evacuation from a nuclear technological disaster. *Geographical Review*, *70*, 1–16. https://doi.org/10.2307/214548

Zhang, Y., Prater, C. S., & Lindell, M. K. (2004). Risk area accuracy and evacuation from Hurricane Bret. *Natural Hazards Review*, *5*, 115–120. https://doi.org/10.1061/(asce)1527-6988(2004)5:3(115)

Zheng, H. (2014). Optimization of bus routing strategies for evacuation. *Journal of Advanced Transportation*, *48*(7), 734–749. https://doi.org/10.1002/atr.1224

Zheng, H., Chiu, Y.-C., Mirchandani, P. B., & Hickman, M. (2010). Modeling of evacuation and background traffic for optimal zone-based vehicle evacuation strategy. *Transportation Research Record: Journal of the Transportation Research Board*, *2196*, 65 74. https://doi.org/10.3141/2196-07

Zhu, Y., Ozbay, K., Xie, K., & Yang, H. (2016). Using big data to study resilience of taxi and subway trips for Hurricanes Sandy and Irene. *Transportation Research Record: Journal of the Transportation Research Board*, *2599*, 70–80. https://doi.org/10.3141/2599-09

SECTION III

Opportunities for New Forms of Organizing during Times of Crisis

Section III of this book takes a more macro-level approach to understand organizing during times of crisis by focusing at the group and community level. This approach moves beyond specific new media to reflect on how different groups organize before, during, and after a crisis. Houston, in Chapter 8 provides an understanding of community resilience, a theme that is echoed in many chapters of this book. To further develop the notion of online communities, Hughes provides an update to an earlier model of online convergence in Chapter 9. By using verbs to describe their organizing actions, this chapter shows how people have drawn upon the affordances of new media to converge in innovative ways. In Chapter 10, Lai invites the reader to consider what happens when a crisis is over and groups that converged are no longer needed. She discusses dormant organizing, a concept that extends beyond her chapter because it was seen in much of the 2017–2018 hurricane activity in the U.S.

FIGURE 0.4 Community organizing for resilience

8

COMMUNITY RESILIENCE AND SOCIAL MEDIA

A Primer on Opportunities to Foster Collective Adaptation Using New Technologies

J. Brian Houston

Resilience is a popular construct in the psychological and social sciences, because it emphasizes the human capacity to experience and recover from challenging events. The idea of resilience is borrowed from the physical sciences, where it represents a material ability to return to form when being stretched or altered. When applied to humans, the idea has been understood to be an ability to *bounce back* from an adverse event (Aldunce, Beilin, Howden, & Handmer, 2014). That is, while a human may experience some distress (e.g., sadness, worry, grief, difficulty sleeping) following a stressful event (e.g., a disaster), resilience is exhibited when an individual recovers and returns to their typical level of functioning (e.g., their usual level of mental and behavioral health) within a time span that is not obviously delayed (Bonanno & Diminich, 2013; Mancini, Bonanno, & Sinan, 2015). In this conceptualization, while humans do bounce *back* to their pre-event level of functioning, they are more accurately bouncing *forward* because they are adjusted to a new reality and experience that has been informed by the challenging event (Houston, 2015). In other words, resilient individuals are not simply returning to *normal* as if nothing happened, they are returning to their usual level of mental and behavioral health, but are likely to be somehow changed because of the experience.

Beyond being applied to individuals (e.g., children, adults), resilience can be applied to any human system (e.g., family, school, neighborhood, community, city, region, nation), which further increases the utility of the construct (Buzzanell & Houston, 2018). This chapter focuses on community resilience specifically, which is typically understood to be a geographic community's ability to recover following a challenge or difficulty. Community resilience can be applied to a variety of issues and events (National Research Council, 2011), but is most frequently applied to community disasters. Community disasters typically include events that are considered natural (e.g., tornadoes, hurricanes, earthquakes) or human-caused. Human-caused disasters may be nonintentional (e.g., chemical spill, transportation accident) or intentional (e.g., terrorist attack, mass shooting). Community resilience is specifically a *collective* activity (Pfefferbaum & Klomp, 2013), in that it requires community members to interact, collaborate, and communicate to facilitate community preparation for and adaption following a challenging event (Acosta, Chandra, & Madrigano, 2017; Brand & Jax, 2007). A community that includes many resilient individuals or resilient organizations, therefore, is not necessarily resilient, as these different entities may not be well connected to help facilitate adaption of the overall community. A variety of community resilience models and frameworks are available in the literature, and several of these are described in the next section.

Community Resilience Models and Frameworks

A popular community resilience model is provided by Norris, Stevens, Pfefferbaum, Wyche, and Pfefferbaum (2008, p. 127), who describe community resilience as involving a "network of adaptive capacities (resources with dynamic attributes)" that allow a community to recover from an event. In this conceptualization, community resilience is considered a process (as opposed to an outcome) that can be observed through a variety of post-event indicators (e.g., community economic productivity, structural rebuilding, mental health).

Norris and colleagues' (2008) community resilience capacities include information and communication, community competence, social capital, and economic development. All of these adaptive capacities in turn include multiple resources. For example, the adaptive capacity of social capital includes social support, social embeddedness, organizational linkages, citizen participation, sense of community, and attachment to place. Overall, the model proposed by Norris and colleagues is robust and provides a variety of areas that communities can consider when attempting to increase their capacity for resilience. At the same time, the numerous areas included in the model can potentially be overwhelming for those seeking to implement community resilience intervention. This complexity is a practical challenge for many conceptualizations of community resilience that are ultimately intended to guide community action.

Pfefferbaum, Pfefferbaum, Nitiéma, Houston, and Van Horn (2015) approach community resilience in a similar fashion as Norris et al. (2008), but instead propose five main domains that include connection and caring, resources, transformative potential, disaster management, and information and communication. These domains can by assessed with community member input (using the Communities Advancing Resilience Toolkit), and results can in turn inform community resilience intervention efforts (Pfefferbaum et al., 2015). Other approaches to community resilience emphasize the community systems and sectors that are important for resilience. For example, Longstaff, Armstrong, Perrin, Parker, and Hidek (2010) describe multiple social subsystems necessary for resilience, including the ecological subsystem, physical infrastructure subsystem, civil society subsystem, and governance subsystem. Gurwitch, Pfefferbaum, Montgomery, Klomp, and Reissman (2007) focus on community sectors for resilience, including media, schools, healthcare, faith-based organizations, and businesses.

Beyond individual models, reviews of community resilience research have identified common themes in the literature. Patel, Rogers, Amlôt, and Rubin (2017) identified several constituent elements that were found to occur across community resilience scholarship. These elements include local knowledge, community networks and relationships, communication, health, governance and leadership, resources, preparedness, and mental outlook. In another review of the community resilience literature, Chandra and colleagues (2010) examined the capabilities frequently included in community resilience definitions and found five common elements. These capability elements include the ability to absorb/resist a disaster, to maintain basic functions during a disaster, to respond to a disaster, to recover after a disaster, and to mitigate threats.

Some community resilience work also include community resilience interventions or actionable frameworks. The Communities Advancing Resilience Toolkit (mentioned previously) includes a survey that can be used for bottom-up community resilience assessment (i.e., assessing citizen perceptions of resilience) and provides resources for community resilience planning and action (Pfefferbaum et al., 2013, 2015). RAND's comprehensive work on community resilience has identified multiple levers of community resilience action that include wellness, access, education, engagement, self-sufficiency, and partnership (Chandra et al., 2011). A variety of tools and resources built around this model are available for communities to use (www.rand.org/multi/resilience-in-action.html) and implementation and evaluation of some of these resources have been reported (Chandra et al., 2013; Plough et al., 2013). Practical approaches to community resilience often include forming community coalitions to lead the effort and guide the work (e.g., resilienceincommunities.com). Lastly, researchers have also approached community resilience from a top-down perspective, in which large data sets (some of which are publicly available) that include indicators theorized to be related to resilience are analyzed to assess the resilience of a geographic area (ranging from low resilience to high resilience). Such approaches have been applied to the U.S. Gulf Coast (Lam, Reams, Li, Li, & Mata, 2016) and the U.S. state of Mississippi (Sherrieb, Norris, & Galea, 2010).

Community Resilience and Communication

Across these various models of and approaches to community resilience, one domain that is generally identified as important is communication. This importance is often explicit in community resilience models. For example, Norris and colleagues' (2008) model includes "information and communication" as one of only four core adaptive capacities, thereby indicating that communication is critical to resilience. Additionally, models often include communicative components that cut across other areas. To again return to the Norris et al. (2008) example, their core capacity of "community competence" includes a variety of activities that are largely communicative (e.g., community problem-solving). More generally, the *collective* nature of community resilience processes implies an essential role for communication (Pfefferbaum & Klomp, 2013). To better articulate the overall role of communication in community resilience, Houston, Spialek, Cox, Greenwood, and First (2015) reviewed the community resilience literature using a communication lens (specifically communication ecology, public relations, and strategic communication perspectives) and developed a framework the conveyed the centrality of communication in community resilience.

Houston, Spialek, and colleagues' (2015) communication and media approach to community resilience includes four main components: communication systems and resources, community relationships, strategic communication processes, and community attributes. Each of these components in turn includes a variety of elements or activities. For example, the communication systems and resources component includes traditional and social media, disaster communication infrastructure (e.g., tornado sirens, emergency alert system), official sources of information (e.g., government agencies), and citizens and organizations. This conceptualization acknowledges that communication related to community resilience can come from top-down (e.g., government agencies providing official disaster warnings) and bottom-up (e.g., citizens talking to each other about a disaster) sources in the community (Houston, 2018). The community relationship component in this model includes interactions and connections such as social support, attachment to place, public–private partnerships, media relations, and more. Community attributes includes characteristics of the community such as the community's flexibility, creativity, diversity, and equality. Finally, strategic communication processes includes several collective sensemaking, problem-solving, and planning processes, such as community planning, community storytelling, disaster and risk planning, economic development discussion, and more. Each of these components can influence the other. So, for example, a community that is diverse and equitable may have different economic development planning and processes than does a community that is homogeneous or that has significant resource disparities between groups in the community. As another example, the media available in a community may influence the overall sense of community or could affect collective community problem-solving.

Community Resilience and Social Media

In this chapter, we utilize Houston, Spialek, et al.'s communication and community resilience model (2015) to consider the role of social media as one specific communication system and resource. The potential utility of social media to affect community resilience also intersects with the functions of social media related to disaster. Social media have uses before, during, and following a disaster (Houston, Hawthorne, et al., 2015), and many of these functions may implicitly affect community resilience. For example, if social media are used to convey disaster warnings, which in turn reduces the impact of an event on individuals (by allowing them to evacuate or make protective preparations), then ultimately this may result in more capacity for resilience, because recovery is more likely due to reduced harm. However, there are also functions of disaster social media that may have more explicit relevance for community resilience. For example, when disaster social media are used to "identify and list ways to help or volunteer, "raise and develop awareness of an event," "donate and receive donations," "provide and receive disaster mental/behavioral health support," "tell and hear stories about the disaster," "discuss socio-political and scientific causes and implications of and responsibility for the events," and "(re) connect community members" (Houston, Hawthorne, et al., 2015, p. 8), then these activities may facilitate adaptation and coping in a community that ultimately results in more resilience related to the event. Overall, when crises occur, individuals tend to exhibit more dependence on media sources as they try to find out what is happening in a threatening situation (Lowrey, 2004), and social media are a place that may be "flooded" with posts as users trying to make sense of the situation (Leykin, Lahad, & Aharonson-Daniel, 2018). As such, in the following section we consider how social media might be used specifically to facilitate strategic communication processes and community relationships that have implications for community resilience (see Figure 8.1).

Social Media Strategic Communication Processes for Community Resilience

As discussed previously, a variety of strategic communication processes can occur in a community to facilitate resilience. These include community competence processes (community planning, action, reflection, problem-solving, and empowerment), community narratives (community storytelling and visioning), community and economic development (development discussion and planning and examination of economic inequities), and disaster and risk processes (disaster and risk information dissemination and use, education, discussion, preparedness, and planning; disaster response coordination; and community resilience awareness; Houston, Spialek et al., 2015). These processes may occur via government agencies, community organizations, and with the involvement of citizens. Within community collaboratives for resilience, social

FIGURE 8.1 Overview of social media functions for fostering community resilience

media may be used in a variety of roles relative to strategic communication processes. Several of these are described below.

Facilitate Meetings

Social media can be used to promote, conduct, expand, and inform community meetings related to resilience. When community resilience meetings occur, social media can be used as a low-cost mechanism to promote and advertise those meetings. Facebook posts and tweets providing details (purpose, time, location) about meetings can be posted and shared among members of the community. Additionally, social media can provide an opportunity to participate in a meeting for individuals who cannot attend in person. For example,

Facebook Live could be used to provide real-time video of a meeting that allows users to provide reactions and comments to the meeting as it occurs. These user comments can be integrated into the meeting discussion in real time. Thus, this use of social media can potentially expand the number of individuals who are able to participate in the events. Additionally, before and after community resilience meetings, social media can be used to gather community input. For example, a social media commenting period might be established and promoted during a period before and after community meetings. Such a process could expand the number of individuals who are able to comment on resilience issues, and provides an opportunity for community members who might be less comfortable providing comments in a public meeting to still contribute to their input.

Share Information and Raise Awareness

Social media represent a low-cost way to share information related to community resilience efforts. Sharing information has been identified as a core use of social media among community organizations (Lovejoy & Saxton, 2012). As previously discussed, the information shared may be about community resilience meetings and planning. Or the information may be risk-related information related to community threats or crises that can inform individual or community action. The information could also address resources, activities, and programs available in the community that may increase resilience. Much of the community resilience information shared may be intended to help raise community awareness about challenges and opportunities in the resilience space.

Promote Action

Communities may utilize social media to promote action or activities related to resilience. Promoted actions may involve advertising community meetings and events, encouraging the use of resources and programs, advocating for policy change, seeking donations and volunteers, and asking for social media support of community resilience efforts (e.g., encouraging community members to share tweets and posts with others). Using social media to take action is another core use of community organizations in general (Lovejoy & Saxton, 2012), and may be important in resilience activities.

Tell and Hear Stories

Social media may be a useful tool for communities to tell and hear resilience stories. These stories may be told and heard organically, through the normal postings on social media and community experiences or events. Or the telling of resilience stories may be facilitated by community organizations or agencies that implement specific resilience activities or programs. As an example, *Picturing Resilience Intervention* (First, Mills-Sandoval, First, & Houston, 2016) is a manual that can be used to help youth tell community resilience stories through images and text, and these results can be shared via social media. Overall, storytelling can be important in fostering community belonging (Ball-Rokeach, Kim, & Matei, 2001; Kim & Ball-Rokeach, 2006), which may help with resilience. In a study of citizen disaster communication, Spialek and Houston (2019) found that more citizen storytelling was related to greater perceptions of community resilience among citizens, providing initial evidence that community storytelling and resilience are linked.

Social Media Relationship-Building for Community Resilience

Social media have the potential to help create and strengthen the community relationships needed for resilience. Relationships that may be beneficial for community resilience include partnerships among citizens, community organizations, political entities, private businesses, and media (Houston, Spialek, et al., 2015). Additionally, citizen connection with and support for each other, as well as a sense of community and community engagement, and attachment to place are relational dimensions that can facilitate resilience. Social media have the potential to establish and strengthen many of these relationships and several opportunities in this area are described below.

Develop and Strengthen Connections

Social media provide an opportunity for users to develop public profiles, follow other users, post content, and read content from others. These capabilities allow connections with other individuals and organizations to be established and strengthened. In this way, online social networks can be established that are similar to and potentially different from existing in-person social networks. In the context of disasters, connecting with others is an important function of citizen disaster communication during an event (Spialek & Houston, 2018).

Build Community and Organize Groups

Beyond developing connections, the relationships possible via social media may be strategically organized to build communities and groups (Afzalan & Evans-Cowley, 2015; Fredericks & Foth, 2013; Lovejoy & Saxton, 2012; Messeter, 2015). Social media tools such as Nextdoor or Facebook may help establish or strengthen connection within a geographic community (e.g., a neighborhood) that could foster opportunities for local resilience awareness and action.

Additionally, interest groups within geographic communities might be formed and sustained via social media, and these groups (e.g., volunteers, environmental activists) could engage in efforts that work to increase the resilience of an area.

Provide Support

Individuals may provide support to each other during or after a disaster using the connections available via social media (Simon, Goldberg, Aharonson-Daniel, Leykin, & Adini, 2014). The availability of social support has repeatedly been found to help individuals cope with disasters (Bonanno, Brewin, Kaniasty, & La Greca, 2010), and therefore facilitating the provision of this support is an important way that social media may foster resilience. Support may include communicating with others to share or find out what happened, providing or receiving emotional support, or giving or receiving practical support or assistance (Houston, Hawthorne, et al., 2015; Tandoc & Takahashi, 2016). Individuals affected by a disaster may seek opportunities to discuss what happened (Houston & Franken, 2015) and social media can provide that space. In a study following tornadoes in the U.S. state of Illinois, Spialek, Czlapinski, and Houston (2016) found that individuals who more frequently used social media to discuss the disaster reported their community to be more connected and caring and have more transformative potential (two community resilience domains). Thus, more opportunity to connect via social media may be related to greater perceptions of community resilience. Relatedly, in a study of flooding in the U.S. state of Colorado, researchers found that perceived information deficits in the community were related to lower perceptions of community resilience, and the deficit of most concern to community members was disconnection from each other (Arneson, Deniz, Javernick-Will, Liel, & Dashti, 2017). Therefore, if social media could grow connections it might foster overall perceptions of community resilience.

Challenges for Social Media and Community Resilience

This chapter has described several ways that social media can be used to foster community resilience via strategic communication and relationship-building pathways. Beyond these opportunities to foster community resilience, several challenges should be noted and are discussed below.

Rumors

During and after disasters, social media content may include untrue rumors about events and issues, and exposure to these rumors can be distressing for individuals (Jones, Thompson, Dunkel Schetter, & Silver, 2017). Rumors on social media may also work against resilience capacities or efforts. For example, rumors about local government or organizations doing something that they are not actually doing (e.g., making a disaster worse through some action) could decrease citizen perceptions of community resilience, by creating mistrust of groups in the community or the government. Therefore, efforts to utilize social media for resilience work will likely need to monitor social media content for rumors or misinformation and intervene to correct information when necessary.

Polarization

Social media may contribute to political polarization (Sunstein, 2017), a process of increasingly divergent sociopolitical viewpoints and attitudes (Fiorina & Abrams, 2008). The challenge of political polarization is that it may make it more difficult for political parties or citizens to find common ground to solve problems. Thus, in the resilience space if polarization prevents compromise and consensus-building around issues of preparing for, responding to, and recovering from community events such as disasters, then this could inhibit community resilience. Social media may strengthen political polarization because the high-choice environment of social media may allow citizens to avoid information they disagree with and seek out information that aligns with their views. These avoidance and reinforcement processes may strengthen existing attitudes, and at the same time, the social media sources providing the reinforcing information may become even more extreme. Therefore, political polarization on social media may be an obstacle for some efforts to incorporate citizen and organizational voices into community resilience work. In particular, because more active social media users may be more polarized, any effort to collect citizen voices via social media to inform community resilience work should include other means of gathering community input, to avoid basing decisions only on social media perspectives that may be more polarized than what is found in the overall community (Hong & Kim, 2016).

Inequality

Inequality is a challenge for social media and community resilience in two main ways. First, the amount of access to and use of social media may vary among groups across a community. For example, older adults may utilize social media less often than younger adults. Despite a general increase in access to online and social media technologies overall, disparities in new technology use and access can mirror other social inequalities (van Deursen & van Dijk, 2014). Therefore, when community resilience efforts use social media, those activities may not be accessible to all members of a community. Ultimately, community resilience efforts should utilize a variety of online and in-person activities and programs, and decisions about what modalities to use should be informed by an understanding of community social media accessibility gaps.

Second, while a large body of community resilience work has been described in this chapter, few of these efforts have directly considered the "historical and structural inequalities" that are impediments to resilience (Acosta et al., 2017, p. 6). Therefore, broadly speaking community resilience planning and programs need to directly address these inequalities, and disparities must be considered specifically in the social media components of community resilience work. On one hand, the potential for the political polarization and access inequality that may occur on social media could be understood to work against progress in the area of improving ongoing community resilience. On the other hand, the connections that are possible via social media could perhaps provide the (mediated) interpersonal interactions that can potentially reduce biases and prejudices (Dixon, Durrheim, & Tredoux, 2005), thus providing space to reduce inequalities through community action and policy. Work is needed to see if the local community resilience context provides a venue for more positive impacts of social media interactions related to sociopolitical inequalities.

Conclusion

Community resilience is the ability of a community to *bounce forward* after a stressful or challenging event such as a disaster (Houston, 2015). While conceptualizations of the capacities and elements that contribute to a community being resilient are often complex, communication is an important factor that is explicitly and implicitly addressed in these models. With regard to communication, community resilience efforts may occur via a variety of communication systems and resources. This chapter reviewed opportunities for using one particular communication source for resilience efforts: social media. Social media may have particular utility in the strategic communication and relationship-building processed related to community resilience (Houston, Spialek, et al., 2015). Strategic communication processes suited for social media include facilitating community resilience meetings, sharing information, raising awareness, promoting action, and telling and hearing stories. Relationship-building processes that may align with social media include developing connections, building community, organizing groups, and providing support. In addition to opportunities to foster community resilience using social media, several challenges exist, including the existence of rumors, polarization, and disparities. Future work should begin to test processes, activities, and interventions that utilize this theoretical framework to determine how social media efforts can be best operationalized and designed to foster more resilient communities in the future.

Practical Implications

This chapter describes several ways that citizens, journalists, organizations, governmental agencies, and communities can potentially utilize social media to facilitate community resilience efforts. The opportunities to use social media to foster community resilience work are practical and can be grouped into strategic communication and relationship-building initiatives. Community resilience interventions and programs are likely to involve conducting meetings, sharing information and stories, establishing and sustaining relationships, organizing action, and providing support. Social media offer low-cost and accessible communication technologies that can help conduct these types of processes. However, social media are not a magic technology that will instantly make a community more resilient. Rather, social media can be used to augment the communication and connections that are necessary for a community to become more resilient. Practical challenges also exist, in that rumors, polarization, and inequities in the social media space can hinder community resilience efforts. Therefore, individuals, organizations, and communities who utilize social media as part of resilience work must be aware of and ready to respond to these potential obstacles.

Discussion Questions

1. Are there other ways to use social media to foster community resilience not described in this chapter?
2. If you were tasked with using social media to improve your community's resilience, where would you start?
3. How might inequalities in social media access be addressed as part of community resilience programs?
4. How could political polarization on social media pose an obstacle to resilience efforts in your community? What could you do to address this obstacle?

References

Acosta, J., Chandra, A., & Madrigano, J. (2017, February). *An agenda to advance integrative resilience research and practice: Key themes from a resilience roundtable.* Santa Monica, CA: RAND.

Afzalan, N., & Evans-Cowley, J. (2015). Planning and social media: Facebook for planning at the neighbourhood scale. *Planning Practice & Research, 30*, 270–285. https://dx.doi.org/10.1080/02697459.2015.1052943

Aldunce, P., Beilin, R., Howden, M., & Handmer, J. (2014). Framing disaster resilience: The implications of the diverse conceptualisations of "bouncing back." *Disaster Prevention and Management: An International Journal, 23*, 252–270. https://dx.doi.org/10.1108/DPM-07-2013-0130

Arneson, E., Deniz, D., Javernick-Will, A., Liel, A., & Dashti, S. (2017). Information deficits and community disaster resilience. *Natural Hazards Review, 18.* https://dx. doi.org/10.1061/(ASCE)NH.1527-6996.0000251

Ball-Rokeach, S. J., Kim, Y. C., & Matei, S. (2001). Storytelling neighborhood: Paths to belonging in diverse urban environments. *Communication Research, 12,* 485–510. https://dx.doi.org/10.1177/009365001028004003

Bonanno, G. A., Brewin, C. R., Kaniasty, K., & La Greca, A. M. (2010). Weighing the costs of disaster: Consequences, risks, and resilience in individuals, families, and communities. *Psychological Science in the Public Interest, 11,* 1–49. https://dx.doi. org/10.1177/1529100610387086

Bonanno, G. A., & Diminich, E. D. (2013). Annual research review: Positive adjustment to adversity—trajectories of minimal-impact resilience and emergent resilience. *Journal of Child Psychology and Psychiatry, 54,* 378–401. https://doi.org/10.1111/ jcpp.12021

Brand, F. S., & Jax, K. (2007). Focusing the meaning(s) of resilience: Resilience as a descriptive concept and a boundary object. *Ecology and Society, 12,* 23. www.ecology andsociety.org/vol12/iss1/art23/

Buzzanell, P. M., & Houston, J. B. (2018). Communication and resilience: Multilevel applications and insights—a *Journal of Applied Communication Research* forum. *Journal of Applied Communication Research, 46,* 1–4. https://dx.doi.org/10.1080/00909882. 2017.1412086

Chandra, A., Acosta, J. D., Howard, S., Uscher-Pines, L., Williams, M. W., Yeung, J. G., & Meredith, L. S. (2011). *Building community resilience to disasters: A way forward to enhance national health security.* Santa Monica, CA: RAND. Retrieved from: www. rand.org/pubs/technical_reports/TR915.html

Chandra, A., Acosta, J., Meredith, L. S., Sanches, K., Stern, S., Uscher-Pines, L., Williams, M., & Yeung, D. (2010). *Understanding community resilience in the context of national health security: A literature review.* Santa Monica, CA: RAND. Retrieved from: www.rand.org/content/dam/rand/pubs/working_papers/2010/RAND_WR 737.pdf

Chandra, A., Williams, M., Plough, A., Stayton, A., Wells, K. B., Horta, M., & Tang, J. (2013). Getting actionable about community resilience: The Los Angeles County Community Disaster Resilience project. *American Journal of Public Health, 103,* 1181–1189. https://dx.doi.org/10.2105/AJPH.2013.301270

Dixon, J., Durrheim, K., & Tredoux, C. (2005). Beyond the optimal contact strategy: A reality check for the contact hypothesis. *American Psychologist, 60,* 697–711. https://dx.doi.org/10.1037/0003-066X.60.7.697

Fiorina, M. P., & Abrams, S. J. (2008). Political polarization in the American public. *Annual Review of Political Science, 11,* 563–588. https://dx.doi.org/10.1146/annurev. polisci.11.053106.153836

First, J., Mills-Sandoval, T., First, N. L., & Houston, J. B. (2016). *Picturing resilience intervention: Using photovoice for youth resilience.* Columbia: University of Missouri Disaster and Community Crisis Center. Retrieved from: https://dcc.missouri.edu/ doc/DCC_2016_Picturing_Resilience_revised.pdf

Fredericks, J., & Foth, M. (2013). Augmenting public participation: Enhancing planning outcomes through the use of social media and Web 2.0. *Australian Planner, 50,* 244–256. https://dx.doi.org/10.1080/07293682.2012.748083

Gurwitch, R. H., Pfefferbaum, B., Montgomery, J., Klomp, R. W., & Reissman, D. B. (2007). *Building community resilience for children and families*. Oklahoma City: Terrorism and Disaster Center at the University of Oklahoma Health Sciences Center.

Hong, S., & Kim, S. H. (2016). Political polarization on Twitter: Implications for the use of social media in digital governments. *Government Information Quarterly, 33*, 777–782. https://dx.doi.org/10.1016/j.giq.2016.04.007

Houston, J. B. (2015). Bouncing forward: Assessing advances in community resilience assessment, intervention, and theory to guide future work. *American Behavioral Scientist, 59*, 175–180. https://dx.doi.org/10.1177/0002764214550294

Houston, J. B. (2018). Community resilience and communication: dynamic interconnections between and among individuals, families, and organizations. *Journal of Applied Communication Research, 46*, 19–22. https://dx.doi.org/10.1080/00909882.2018.1426704

Houston, J. B., & Franken, N. J. (2015). Disaster interpersonal communication and posttraumatic stress following the 2011 Joplin, Missouri tornado. *Journal of Loss and Trauma, 20*, 195–206. https://dx.doi.org/10.1080/15325024.2013.848614

Houston, J. B., Hawthorne, J., Perreault, M. F., Park, E. H., Goldstein Hode, M., Halliwell, M. R., Turner McGowen, S. E., Davis, R., Vaid, S., McElderry, J. A., & Griffith, S. A. (2015). Social media and disasters: A functional framework for social media use in disaster planning, response, and research. *Disasters, 39*, 1–22. https://dx.doi.org/10.1111/disa.12092

Houston, J. B., Spialek, M. L., Cox, J., Greenwood, M. M., & First, J. (2015). The centrality of communication and media in fostering community resilience: A framework for assessment and intervention. *American Behavioral Scientist, 59*, 270–283. https://dx.doi.org/10.1177/0002764214548563

Houston, J. B., Spialek, M. L., First, J., Stevens, J., & First, N. L. (2017). Individual perceptions of community resilience following the 2011 Joplin tornado. *Journal of Contingencies and Crisis Management, 25*, 354–363. https://dx.doi.org/10.1111/1468-5973.12171

Jones, N. M., Thompson, R. R., Dunkel Schetter, C., & Silver, R. C. (2017). Distress and rumor exposure on social media during a campus lockdown. *Proceedings of the National Academy of Sciences, 114*, 11663–11668. https://dx.doi.org/10.1073/pnas.1708518114

Kim, Y. C., & Ball-Rokeach, S. J. (2006). Civic engagement from a communication infrastructure perspective. *Communication Theory, 16*, 1–25. https://dx.doi.org/10.1111/j.1468-2885.2006.00267.x

Lam, N. N., Reams, M., Li, K., Li, C., & Mata, L. P. (2016). Measuring community resilience to coastal hazards along the northern gulf of Mexico. *Natural Hazards Review, 17*, 04015013. https://dx.doi.org/10.1061/(ASCE)NH.1527-6996.0000193

Leykin, D., Lahad, M., & Aharonson-Daniel, L. (2018). Gauging urban resilience from social media. *International Journal of Disaster Risk Reduction, 31*, 393–402. https://dx.doi.org/10.1016/j.ijdrr.2018.04.021

Longstaff, P. H., Armstrong, N. J., Perrin, K. A., Parker, W. M., & Hidek, M. (2010). Building resilient communities: A preliminary framework for assessment. *Homeland Security Affairs, 6*, 1–23.

Lovejoy, K., & Saxton, G. D. (2012). Information, community, and action: How nonprofit organizations use social media. *Journal of Computer-Mediated Communication, 17*, 337–353. https://dx.doi.org/10.1111/j.1083-6101.2012.01576.x

Lowrey, W. (2004). Media dependency during a large-scale social disruption: The case of September 11. *Mass Communication & Society, 7*, 339–357.

Mancini, A. D., Bonanno, G. A., & Sinan, B. (2015). A brief retrospective method for identifying longitudinal trajectories of adjustment following acute stress. *Assessment, 22*, 298–308. https://dx.doi.org/10.1177/1073191114550816

Messeter, J. (2015). Social media use as urban acupuncture for empowering socially challenged communities. *Journal of Urban Technology, 22*, 79–96. https://dx.doi.org/10.1080/10630732.2015.1040291

National Research Council. (2011). *Building community disaster resilience through private–public collaboration.* Washington, DC: National Academies Press.

Norris, F. H., Stevens, S. P., Pfefferbaum, B., Wyche, K. F., & Pfefferbaum, R. L. (2008). Community resilience as a metaphor, theory, set of capacities, and strategy for disaster readiness. *American Journal of Community Psychology, 41*, 127–150.

Patel, S. S., Rogers, M. B., Amlôt, R., & Rubin, G. J. (2017). What do we mean by "community resilience"? A systematic literature review of how it is defined in the literature. *PLoS Currents, 9.* http://currents.plos.org/disasters/index.html%3Fp=28783.html

Pfefferbaum, R. L., & Klomp, R. W. (2013). Community resilience, disasters, and the public's health. In F. G. Murphy (Ed.), *Community engagement, organization, and development for public health practice* (pp. 275–298). New York, NY: Springer.

Pfefferbaum, R. L., Pfefferbaum, B., Nitiéma, P., Houston, J. B., & Van Horn, R. L. (2015). Assessing community resilience: An application of the expanded CART survey instrument with affiliated volunteer responders. *American Behavioral Scientist, 59*, 181–199. https://dx.doi.org/10.1177/0002764214550295

Pfefferbaum, R. L., Pfefferbaum, B., Van Horn, R. L., Klomp, R. W., Norris, F. H., & Reissman, D. B. (2013). The Communities Advancing Resilience Toolkit (CART): An intervention to build community resilience to disasters. *Journal of Public Health Management and Practice, 19*, 250–258. https://dx.doi.org/10.1097/PHH.0b013e318268aed8

Plough, A., Fielding, J. E., Chandra, A., Williams, M., Eisenman, D., Wells, K. B., Law, G. Y., Fogleman, S., & Magaña, A. (2013). Building community disaster resilience: Perspectives from a large urban county department of public health. *American Journal of Public Health, 103*, 1–8. https://dx.doi.org/10.2105/AJPH.2013.301268

Sherrieb, K., Norris, F., & Galea, S. (2010). Measuring capacities for community resilience. *Social Indicators Research, 99*, 227–247.

Simon, T., Goldberg, A., Aharonson-Daniel, L., Leykin, D., & Adini, B. (2014). Twitter in the cross fire—the use of social media in the Westgate Mall terror attack in Kenya. *PLoS One, 9*, e104136. https://dx.doi.org/10.1371/journal.pone.0104136

Spialek, M. L., Czlapinski, H. M., & Houston, J. B. (2016). Disaster communication ecology and community resilience perceptions following the 2013 Central Illinois tornadoes. *International Journal of Disaster Risk Reduction, 17*, 154–160. https://dx.doi.org/10.1016/j.ijdrr.2016.04.006

Spialek, M. L., & Houston, J. B. (2018). The development and initial validation of the citizen disaster communication assessment. *Communication Research, 45*, 934–955. https://dx.doi.org/10.1177/0093650217697521

Spialek, M. L., & Houston, J. B. (2019). The influence of citizen disaster communication on perceptions of neighborhood belonging and community resilience. *Journal of Applied Communication Research, 47*(1), 1–23. doi:10.1080/00909882.2018.1544718

Sunstein, C.R. (2017). *#republic: Divided democracy in the age of social media*. Princeton, NJ: Princeton University Press.

Tandoc, E. C., & Takahashi, B. (2016). Log in if you survived: Collective coping on social media in the aftermath of Typhoon Haiyan in the Philippines. *New Media & Society, 19*, 1778–1793. https://dx.doi.org/10.1177/1461444816642755

van Deursen, A. J. A. M., & van Dijk, J. A. G. M. (2014). The digital divide shifts to differences in usage. *New Media & Society, 16*, 507–526. https://dx.doi.org/10.11 77/1461444813487959

9

SITE-SEEING IN DISASTER

Revisiting Online Social Convergence a Decade Later

Amanda Lee Hughes

In 2008, my colleagues and I wrote a paper about how people come together, or converge, online around disaster events (Hughes, Palen, Sutton, Liu, & Vieweg, 2008). Sociologists have long documented how people physically converge on disaster sites to help the response effort, mourn, and learn more about the event (Fritz & Mathewson, 1957; Kendra & Wachtendorf, 2003), and this paper drew parallels between physical and online convergence behaviors. Social media were new at the time and researchers were only beginning to investigate and understand the types of interactions and behaviors that social media could support. Thus, our investigation was preliminary and fledgling.

Since that time, a large and growing body of research has explored how people use social media during disaster events (Palen & Hughes, 2018). Most of this research falls in the field of *crisis informatics* (Hagar & Haythornthwaite, 2005; Palen, Vieweg, Liu, & Hughes, 2009), where researchers critically examine the complex socio-technical information environment that surrounds a disaster or crisis event. This chapter revisits online convergence in disaster with fresh perspective, drawing from over a decade of crisis informatics research. The chapter focuses on what we have learned about online convergence behavior and how it has changed (and continues to change) the way we respond to disaster events.

Convergence Framework

The convergence framework used in this chapter originated from Fritz and Mathewson (1957) and was later expanded by Kendra and Wachtendorf (2003). This framework identifies seven types of people that physically converge on a disaster site: helpers, the anxious, returnees, supporters, mourners, exploiters, and the curious. Like our original analysis of online convergence (Hughes et al., 2008), this chapter uses verbs to describe the converger types: helping, being anxious, returning, supporting, mourning, exploiting, and being curious. Verbs are used because it is difficult to tie convergent actions and behaviors to particular people in the online space as can be done more readily in a physical environment. This convergence framework serves as a useful tool to understand how people use online media during a crisis event because it helps us better explore the different types of convergence behaviors and identify potential gaps in our knowledge. It also allows us to draw parallels between behavior found in the physical and online world and potentially apply findings from each environment to the other. In the following sections, this chapter reconsiders each of the seven types of online convergence behavior in turn, with an emphasis on new and/or changing behaviors and phenomenon.

Helping

> Helpers converge in order to assist victims or responders.
>
> *(Kendra & Wachtendorf, 2003, p. 107)*

As a disaster event unfolds, members of the public use online channels to aid victims of the disaster as well as emergency responders. For example, people have used social media to coordinate humanitarian relief efforts, identify those in need of assistance, and organize online fundraising efforts for disaster victims (Okada, Ishida, & Yamauchi, 2017; Peary, Shaw, & Takeuchi, 2012; Starbird & Palen, 2011). Many of these offers of help are spontaneously given by individuals as they recognize needs, but recent years have seen these helping activities move towards more organized behavior (Schmidt, Wolbers, Ferguson, & Boersma, 2017; White & Palen, 2015).

A more formalized class of helpers has emerged since our original analysis of online convergence behavior called *digital volunteers* (Starbird & Palen, 2011). These volunteers organize themselves around disaster needs and use computing resources to meet those needs. Because digital volunteers use online computing resources, they can be located virtually anywhere in the world. This global distribution of digital volunteers can facilitate disaster resilience because help can come from those who are remote and not physically affected by the disaster event.

An example of digital volunteers are crisis mappers, who create and maintain digital maps during disaster events (Norheim-Hagtun & Meier, 2010; Soden & Palen, 2014). Crisis mappers provide a valuable service because current and accurate maps during a disaster event are often difficult to find and produce, especially when the physical environment may be rapidly changing (like during an earthquake or a wildfire). Mappers combine open-source spatial data with other information sources about the disaster event (e.g., shelter locations, road closures, fire perimeters, photos, and videos) to create interactive, dynamic digital maps that facilitate response and relief efforts (Soden & Palen, 2016).

Nonprofit organizations are also involved in digital volunteerism. One organized group, Humanity Road, responds to events around the world by matching online requests to available resources (Starbird & Palen, 2013). Other research and nonprofit groups are developing software to collect and analyze citizen-generated information (Castillo, 2016; Imran, Castillo, Lucas, Meier, & Vieweg, 2014; Meier & Brodock, 2008). For example, the AIDR (Artificial Intelligence for Disaster Response) system is a platform where humans inform machine learning algorithms by classifying social media messages as informative or non-informative (Imran et al., 2014). The system learns how to recognize informative disaster messages more quickly and with more accuracy using real-time human input. Another system built by Purohit and colleagues (2014) parses social media streams using machine classification techniques to identify people who need assistance during a disaster and those who can meet those needs. While we have seen many advances in using machines to analyze large amounts of social media data, the challenge of extracting only the useful and actionable information from social media and presenting it in a digestible way remains a challenge that practitioners and researchers continue to work on (Palen & Hughes, 2018).

The term Virtual Operations Support Teams (VOSTs) has emerged to describe a group of trusted digital volunteers (many with prior emergency management experience) who remotely assist with monitoring and distributing information through social media (St. Denis, Hughes, & Palen, 2012; St. Denis, Palen, & Anderson, 2014). The role of a VOST is to support on-site emergency responders in managing the often-overwhelming amount of online data generated during a disaster. Examples of tasks that members of a VOST might perform include posting crisis information to official response sites, relaying social media information that could be useful to the official responders, and identifying and correcting false rumor or misinformation. The VOST concept started informally, but over time organizations (such as the Virtual Operation Support Group[1]) have been created to more formally define the role of a VOST, its operating procedures, and best practices. VOSTs currently operate in locations around the world and can provide much-needed expertise and assistance during disaster events when resources are low.

Challenges to Using Digital Volunteers

While digital volunteers have proven useful, there are many challenges to incorporating them into official response efforts (Hughes & Tapia, 2015). Many digital volunteers are part of grassroots efforts that emerge as a disaster event unfolds. As such, these volunteers are unlikely to have pre-existing connections with emergency response organizations and thus they have no way to feed their efforts into these organizations. Similarly, emergency response organizations may not know what services digital volunteers can offer or they may not trust the information they provide, so these volunteers are unlikely to be used as effectively as they could be (Hughes & Tapia, 2015). Also, the work of digital volunteers are often duplicated because there is no central body organizing and coordinating them (Hughes & Tapia, 2015). Just as there are challenges with incorporating volunteers who physically converge on the site of a disaster into formal response efforts (e.g., well-intentioned people who donate unneeded goods that then must be managed), we see the same kinds of issues in the online world.

In summary, online helping behavior has significantly grown in scope and scale over the past decade. Through social media and other online tools, everyday citizens have found ways to contribute to disaster relief efforts, even when located at a distance. These online relief efforts have become increasingly organized, as people seek to optimize their contributions. Yet, there remain significant challenges to leveraging this helpful online behavior, mostly centered on how to meaningfully incorporate digital volunteer work into formal emergency response efforts.

Being Anxious

> The anxious are people from outside the impacted area who attempt to obtain information about family and friends.
>
> *(Kendra & Wachtendorf, 2003, p. 105)*

Following any disaster event, anxious people attempt to understand how the event has affected them and their social network. The prior online convergence paper (Hughes et al., 2008) noted the existence of applications to report the status of one's wellbeing during a disaster (e.g., the Safe and Well[2] service from the American Red Cross). These applications have since become more common and sophisticated. Companies, such as Facebook,[3] have built features into their platform that let people report their own status as well as check on the status of those in their network.

A growing number of online tools, like social media, also support collective intelligence as anxious people seek to understand disaster events and their impact. The original online convergence paper (Hughes et al., 2008) described how people used online tools to assess the impact of the April 16, 2007 Virginia Tech school shooting on their social circle. Together members of the affected community accurately discovered the names of all the deceased before the names were officially released the following day (Palen et al., 2009; Vieweg, Palen, Liu, Hughes, & Sutton, 2008).

This kind of collective intelligence behavior continues to today. For example, Wikipedia is a site where information about disaster events is collectively created and curated as people make sense of and collectively report on these events (Keegan, 2015; Keegan, Gergle, & Contractor, 2013). A social media discussion site called Reddit is another popular place for the anxious to find information about a disaster event as it progresses. Researchers have found that as Reddit users make information more or less visible, it contributes to shaping the narrative of the event (Leavitt & Clark, 2014; Leavitt & Robinson, 2017). A popular online forum in China—called Tianya—served as a place where people could share and seek information following the 2008 Sichuan earthquake (Qu, Wu, & Wang, 2009). These examples are only a few of the many online platforms and sites where the anxious have sought information about disaster events that concern them.

In the past ten years, the growing number of online information sources has driven expectations around how quickly the anxious expect to find and receive information about a disaster event (Stephens & Malone, 2009). Through online media, people can find information from their peers in real time. Consequently, emergency responders and the news media find that they must distribute disaster information more quickly, otherwise people will seek information elsewhere (Hughes & Palen, 2012; Stephens & Malone, 2009). These changes have resulted in faster sharing of information, often with little to no vetting, which has contributed to an increasing concern about the spread of misinformation during a disaster event (Huang, Starbird, Orand, Stanek, & Pedersen, 2015; Oyeyemi, Gabarron, & Wynn, 2014).

Returning

> Returnees in New York City [during the World Trade Center Attacks] included residents, employees, business owners, and "substitute" returnees (Fritz and Mathewson, 1957) who are the relatives and friends of disaster victims who enter the area to assess the victims' losses and salvage their property.
>
> *(Kendra & Wachtendorf, 2003, p. 104)*

Through innovation in online tools and services, the past decade has seen many new ways in which people can virtually return to the disaster site. The use of digital imagery, especially with improved unmanned aerial vehicle (UAV) technology, has made affordable, safe, and near real-time inspection of disaster sites possible (Adams, Levitan, & Friedland, 2012). In addition, citizen journalists (Gillmor, 2006), armed with digital devices, can act as sensors and report information from the site to those who are remotely located (Goodchild, 2007). These reports may include video, photos, and/or textual content. Often the first reports from a disaster site come from citizens and not the traditional news media. Applications, such as Ushahidi (Meier & Brodock, 2008; Morrow, Mock, Papendieck, & Kocmich, 2011) attempt to map and make sense of the data that citizens provide through online media. Such applications can help emergency responders as well as those directly affected by the event to virtually return through the use of technology to the physical disaster site and assess the impact. For example, one study determined that social media data could be used to discover spatial flood patterns and the depths of flood waters quickly and without having to physically deploy to the location (Fohringer, Dransch, Kreibich, & Schröter, 2015). In another example, Dashti et al. (2014) found that images and videos shared through social media could help geotechnical experts collect reconnaissance data on disaster-affected infrastructure (e.g., bridges, roads, buildings) to assess failures and plan ways to minimize future failures.

Supporting

> The convergers are either individuals or groups who gathered to encourage and express gratitude to emergency workers.
>
> *(Kendra & Wachtendorf, 2003, p. 115)*

Much of the activity found online during disaster events is supportive. For instance, researchers have found that many online messages during an event contain condolences, prayers, support, and expressions of thanks (Olteanu, Vieweg, & Castillo, 2015; Shaw, Burgess, Crawford, & Bruns, 2013). A common convention in social media sites is to change one's profile picture to show support for the victims of a disaster event. Another example is found in the recent growth of crowdfunding websites where users can create and promote custom donation sites that collect money for different disaster needs (Kuppuswamy & Bayus, 2018). Other research has shown that social media can help people establish social support and solidarity during political protests (Starbird & Palen, 2012; Tonkin, Pfeiffer, & Tourte, 2012), times of war (Mark, Al-Ani, & Semaan, 2009; Mark & Semaan, 2008), and acts of terror (Eriksson, 2016; Glasgow, Vitak, Tausczik, & Fink, 2016).

Supportive online behavior can also be directed toward emergency responders. Many responders maintain their own official social media sites, which gives citizens a direct communication channel to them (Hughes, St. Denis, Palen, & Anderson, 2014). These sites are often platforms where citizens will express gratitude for emergency response efforts. However, the online communication channels that afford these supportive behaviors also afford opportunities for people to criticize or even attack those who they feel are not responding appropriately to a disaster event.

The past decade has seen an increase in the ways that people can support disaster victims and the associated relief effort. Through the Internet, people can offer support (e.g., money, services, words of encouragement) without restriction to time and space. People can also interact more directly with emergency responders through social media. These changes have a democratizing effect in that they allow opportunities for all people to be more involved in disaster management.

Mourning

> These were people who went to locations such as firehouses, Union Square, and Ground Zero to lay flowers, light candles, create memorials, and mourn the dead.
>
> *(Kendra & Wachtendorf, 2003, p. 117)*

Over the past ten years, we have seen an increase in the number of people who come together online to create digital memorials and post online messages to mourn the victims of a disaster. During the Great East Japan earthquake of 2011, residents used social media to share their grief and mourn with their community (Hjorth & Kim, 2011). Researchers have studied several online efforts that curate photos, videos, and written experiences about crisis events so that society can collectively mourn, remember, and learn from these events (Liu, Palen, & Giaccardi, 2012; Mark et al., 2012). In addition, crisis named resources (CNRs) appear following nearly every event. These resources are websites as well as social media accounts and pages that are created in response to, and named after, a disaster or crisis event (Chauhan & Hughes, 2018). Because they are easy to find when searching on the disaster name, these CNRs often serve as online rallying points where people come to support the victims of an event and mourn their losses.

Exploiting

> Exploiters are convergers who use the disaster for personal gain or profit.
>
> *(Kendra & Wachtendorf, 2003, p. 113)*

We noted the existence of exploitive behavior online during our original study, but these behaviors were not seen as frequently as they are today. Fake donation sites that appear shortly after a disaster event are a growing problem (Scott, 2017). These sites are designed to look like authentic donation sites from trusted groups, such as the American Red Cross. For instance, the "Hurricane Sandy Relief Effort" was a fake charity created by con artists that raised $600k, supposedly for victims (Scott, 2017).

Perhaps the biggest development in exploitive behavior over the past decade has been regarding the spread of false rumors and misinformation during crisis events. With the many different ways that people are connected online, unverified information can spread quickly before it can be corroborated and/or corrected. Often the sharing or spread of this information is not intentionally exploitive. There is some evidence that people will self-regulate by questioning and correcting information (Castillo, Mendoza, & Poblete, 2011), but this is not always the case in an increasingly large and complex information environment. In other cases, people intentionally spread misinformation. During Hurricane Sandy, reports that the New York Stock Exchange was underwater and that there were sharks swimming in the front yards of people in New Jersey spread widely over social media (Schulten, 2015). In another example, researchers revealed the ways in which people create alternative narratives for mass shooting events (some even claiming the shootings were staged by actors) that are usually designed to promote different political agendas (Starbird, 2017). Thus, researchers are seeking ways to identify more credible information (Gupta & Kumaraguru, 2012; Thomson et al., 2012) as well as ways to identify false rumors and misinformation so that they can be quickly addressed (Arif et al., 2016; Maddock et al., 2015).

Being Curious

> Curious convergers come to the impacted site primarily to view the destruction left in the wake of the disaster and the activities surrounding the response.
>
> *(Kendra & Wachtendorf, 2003, p. 111)*

Curious behavior is more broadly supported than ten years ago. Through online media, people can observe the progression of an event through shared video, text, and images. These media let more people connect around an event on a larger scale that previously possible and from any distance (Bruns & Burgess, 2012; Palen & Vieweg, 2008). However, there is some concern that broader exposure to so many disaster events may desensitize us as a society (Li, Conathan, & Hughes, 2017).

Online curious behavior continues to be hard to quantify because it often cannot be directly observed. Many people look online for disaster information but they do not necessarily leave traces of their presence (e.g., likes, shares, comments, etc.). However, there is no shortage of information available to those curious about an event, and often the challenge today is to find relevant sources among the vast amount of available information. This challenge remains an open problem, but many researchers are working on algorithms and systems that attempt to sift through the deluge of social media data available around a disaster event to find relevant and actionable information (Abel, Hauff, Houben, & Stronkman, 2012; Imran, Castillo, Diaz, & Vieweg, 2015; Reuter, Marx, & Pipek, 2012).

Discussion

Online media have enabled disaster convergence on a new scale. People from around the world can offer help and support for victims through their computing devices. They can also seek information to better understand the impact of a disaster and how they should respond. Online media have also enabled a new set of digital volunteers that can assist disaster victims in new and unprecedented ways.

When we wrote the original convergence paper (Hughes et al., 2008), the use of online media to find and share information during a disaster event was a relatively new phenomenon. Drawing attention to how people could converge online was a novel academic contribution. Over time, the use of social media and other online tools in disaster has become more commonplace. Online media have become an integral part of the information ecosystem that affects how people respond to and interpret disaster events (De Choudhury, Monroy-Hernández, & Mark, 2014; Monroy-Hernández, boyd, Kiciman, De Choudhury, & Counts, 2013). Thus researchers have turned their attention to deeper concerns, such as the empirical study of online convergence behavior (Chauhan & Hughes, 2018; Kogan, Palen, & Anderson, 2015; Schmidt et al., 2017; Semaan & Mark, 2012; Soden & Palen, 2014; Starbird & Palen, 2011) and the creation of software systems to support that activity (Abel et al., 2012; Hughes & Shah, 2016; Imran et al., 2014; Morrow et al., 2011; Reuter, Ludwig, Kaufhold, & Pipek, 2015). These efforts have been taken up with increasing frequency and are only likely to continue.

This chapter concludes by identifying three of the greatest areas for growth regarding online social convergence research:

1. Making sense of all the online data about a disaster event.
2. Discovering better ways to integrate digital volunteer efforts into formal emergency response.
3. Improving the veracity of online information.

To address the first area for growth, crisis informatics researchers from many disciplines (such as computer science, information science, and data visualization) are dedicated to finding more effective ways to filter, verify, and make sense of large amounts of social media data (Imran et al., 2015). There is also a focus on providing tools to everyday citizens to help them analyze data during a disaster event and make decisions (Palen, Vieweg, & Anderson, 2011). Despite much progress, finding useful information during a disaster event is still an open problem that continues to attract new research.

The second area for growing research interest is in seeking ways to efficiently integrate citizen-generated data and digital volunteer efforts into more formal emergency response organizations (Hughes & Tapia, 2015). With increased opportunities to participate, we have seen shifts in how emergency responders think of the public; there seems to be less of a focus on managing public online behavior, and more of a focus on finding ways to work together during a disaster event. Recognition that digital volunteers can bring value to formal disaster management has spawned a growing number of research- and practitioner-led efforts (Hughes & Tapia, 2015; Palen, Soden, Anderson, & Barrenechea, 2015; St. Denis et al., 2012) to find the most effective ways to collaborate.

Finally, the last area of research growth is related to control, trust, and information veracity. Emergency responders struggle to control the flow of information around a disaster event, as they seek to understand the information that people are sharing and try to mitigate misinformation and false rumors. These responders find that they cannot manage the information that circulates around a disaster event like they could before widespread adoption of Internet-enabled devices (Hughes & Palen, 2012). Similarly, members of the disaster-affected public must sift through an overabundance of online information that may or may not be accurate or trustworthy. They are simply overloaded with information. Concern about the spread of misinformation during critical events has escalated in recent years (Starbird, 2017), which has brought a sense of urgency to the research needed to understand and counter it. Consequently, we have seen a recent increase in the number of studies that address the topic—a trend that is expected to continue.

This chapter summarizes online social convergence behavior and offers perspectives on how it has changed over the past decade. As disasters continue to happen and online platforms continue to grow in use and complexity, it will be increasingly important to understand how people converge online before, during, and after a disaster event. Such understandings will lead to more effective disaster response, by shaping future technology design and informing disaster management policy.

Discussion Questions

After reading this chapter, consider the following discussion questions:

1. How do social media technologies support convergence behaviors?
2. How do social media technologies discourage convergence behaviors?
3. Are there physical convergence behaviors that are not well supported by social media technology?
4. Can online spaces serve as a substitute for physical convergence?
5. How can the online convergence behaviors of the public be most effectively incorporated into formal response efforts?

Notes

1 https://vosg.us/
2 https://safeandwell.communityos.org/cms/index.php
3 Facebook's Safety Check: www.facebook.com/about/crisisresponse/

References

Abel, F., Hauff, C., Houben, G.-J., Stronkman, R., & Tao, K. (2012). Semantics + filtering + search = Twitcident exploring information in social web streams. In *Proceedings of the 23rd ACM Conference on Hypertext and Social Media* (pp. 285–294). New York, NY: ACM Press. https://dx.doi.org/10.1145/2309996.2310043

Adams, S. M., Levitan, M. L., & Friedland, C. J. (2012). High resolution imagery collection utilizing unmanned aerial vehicles (UAVs) for post-disaster studies. *Advances in Hurricane Engineering.* https://dx.doi.org/10.1061/9780784412626.067

Arif, A., Shanahan, K., Chou, F.-J., Dosouto, Y., Starbird, K., & Spiro, E. S. (2016). How information snowballs: Exploring the role of exposure in online rumor propagation. In *Proceedings of the 19th ACM Conference on Computer-Supported Cooperative Work & Social Computing* (pp. 466–477). New York, NY: ACM. https://dx.doi.org/10.1145/2818048.2819964

Bruns, A., & Burgess, J. E. (2012). Local and global responses to disaster: #eqnz and the Christchurch Earthquake. In *Proceedings of the Disaster and Emergency Management Conference* (pp. 86–103). Brisbane, QLD: AST Management Pty Ltd.

Castillo, C. (2016). *Big crisis data: Social media in disasters and time-critical situations.* New York, NY: Cambridge University Press.

Castillo, C., Mendoza, M., & Poblete, B. (2011). Information credibility on Twitter. In *Proceedings of the 20th International Conference on World Wide Web* (pp. 675–684). New York, NY: ACM. https://dx.doi.org/10.1145/1963405.1963500

Chauhan, A., & Hughes, A. L. (2018). Social media resources named after a crisis event. In *Proceedings of the 2018 Information Systems for Crisis Response and Management Conference (ISCRAM 2018).* http://idl.iscram.org/files/apoorvachauhan/2018/1580_ ApoorvaChauhan+AmandaLeeHughes2018.pdf

Dashti, S., Palen, L., Heris, M. P., Anderson, K. M., Anderson, J., & Anderson, S. (2014). Supporting disaster reconnaissance with social media data: A design-oriented case study of the 2013 Colorado floods. In *Proceedings of the Information Systems for Crisis Response and Management Conference (ISCRAM 20014).* University Park, PA. http:// idl.iscram.org/files/dashti/2014/423_Dashti_etal2014.pdf

De Choudhury, M., Monroy-Hernández, A., & Mark, G. (2014). "Narco" emotions: Affect and desensitization in social media during the Mexican Drug War. In *Proceedings of the IGCHI Conference on Human Factors in Computing Systems* (pp. 3563–3572). New York, NY: ACM. https://dx.doi.org/10.1145/2556288.2557197

Eriksson, M. (2016). Managing collective trauma on social media: The role of Twitter after the 2011 Norway Attacks. *Media, Culture & Society, 38*(3), 365–380. https:// dx.doi.org/10.1177/0163443715608259

Fohringer, J., Dransch, D., Kreibich, H., & Schröter, K. (2015). Social media as an information source for rapid flood inundation mapping. *Natural Hazards and Earth System Sciences Discussions, 3,* 4231–4264. https://dx.doi.org/10.5194/nhessd-3-4231-2015

Fritz, C. E., & Mathewson, J. H. (1957). *Convergence behavior in disasters: A problem in social control, committee on disaster studies.* Washington: National Academy of Sciences, National Research Council.

Gillmor, D. (2006). *We the media: Grassroots journalism by the people, for the people.* Sebastopol, CA: O'Reilly Media.

Glasgow, K., Vitak, J., Tausczik, Y., & Fink, C. (2016). Grieving in the 21st century: Social media's role in facilitating supportive exchanges following community-level traumatic events. In *Proceedings of the 7th 2016 International Conference on Social*

Media & Society (pp. 4:1–4:10). New York, NY: ACM. https://dx.doi.org/10.1145/2930971.2930975

Goodchild, M. F. (2007). Citizens as sensors: The world of volunteered geography. *GeoJournal, 69*(4), 211–221. https://dx.doi.org/10.1007/s10708-007-9111-y

Gupta, A., & Kumaraguru, P. (2012). Credibility ranking of tweets during high impact events. In *Proceedings of the 1st Workshop on Privacy and Security in Online Social Media* (pp. 2:2–2:8). New York, NY: ACM Press. https://dx.doi.org/10.1145/2185354.2185356

Hagar, C., & Haythornthwaite, C. (2005). Crisis, farming & community. *The Journal of Community Informatics, 1*(3), 41–52.

Hjorth, L., & Kim, K.-H. Y. (2011). Good grief: The role of social mobile media in the 3.11 earthquake disaster in Japan. *Digital Creativity, 22*(3), 187–199. https://dx.doi.org/10.1080/14626268.2011.604640

Huang, Y. L., Starbird, K., Orand, M., Stanek, S. A., & Pedersen, H. T. (2015). Connected through crisis: Emotional proximity and the spread of misinformation online. In *Proceedings of the 18th ACM Conference on Computer Supported Cooperative Work & Social Computing* (pp. 969–980). New York, NY: ACM. https://dx.doi.org/10.1145/2675133.2675202

Hughes, A. L., & Palen, L. (2012). The evolving role of the public information officer: An examination of social media in emergency management. *Journal of Homeland Security and Emergency Management, 9*(1). https://dx.doi.org/10.1515/1547-7355.1976

Hughes, A. L., Palen, L., Sutton, J., Liu, S. B., & Vieweg, S. (2008). "Site-seeing" in disaster: An examination of on-line social convergence. In *Proceedings of the Information Systems for Crisis Response and Management Conference (ISCRAM 2008)*. Washington, DC. http://idl.iscram.org/files/hughes/2008/605_Hughes_etal2008.pdf

Hughes, A. L., & Shah, R. (2016). Designing an application for social media needs in emergency public information work. In *Proceedings of the 19th International Conference on Supporting Group Work* (pp. 399–408). New York, NY: ACM. https://dx.doi.org/10.1145/2957276.2957307

Hughes, A. L., St. Denis, L. A., Palen, L., & Anderson, K. M. (2014). Online public communications by police & fire services during the 2012 Hurricane Sandy. In *Proceedings of the 2014 International Conference on Human Factors in Computing Systems (CHI 2014)* (pp. 1505–1514). New York, NY: ACM Press. https://dx.doi.org/10.1145/2556288.2557227

Hughes, A. L., & Tapia, A. H. (2015). Social media in crisis: When professional responders meet digital volunteers. *Journal of Homeland Security and Emergency Management, 12*(3), 679–706. https://doi.org/10.1515/jhsem-2014-0080

Imran, M., Castillo, C., Diaz, F., & Vieweg, S. (2015). Processing social media messages in mass emergency: A survey. *ACM Computing Surveys, 47*(4), 67:1–67:38. https://dx.doi.org/10.1145/2771588

Imran, M., Castillo, C., Lucas, J., Meier, P., & Vieweg, S. (2014). AIDR: Artificial intelligence for disaster response. In *Proceedings of the Companion Publication of the 23rd International Conference on World Wide Web Companion* (pp. 159–162). Republic and Canton of Geneva, Switzerland: International World Wide Web Conferences Steering Committee. https://dx.doi.org/10.1145/2567948.2577034

Keegan, B. C. (2015). Emergent social roles in Wikipedia's breaking news collaborations. In E. Bertino & S. A. Matei (Eds.), *Roles, trust, and reputation in social media knowledge markets* (pp. 57–79). New York, NY: Springer.

Keegan, B. C., Gergle, D., & Contractor, N. (2013). Hot off the wiki: Structures and dynamics of Wikipedia's coverage of breaking news events. *American Behavioral Scientist, 57*(5), 595–622. https://dx.doi.org/10.1177/0002764212469367

Kendra, J. M., & Wachtendorf, T. (2003). Reconsidering convergence and converger: Legitimacy in response to the World Trade Center Disaster. In L. Clarke (Ed.), *Terrorism and disaster: New threats, new ideas* (pp. 97–122). Amsterdam: Emerald Group.

Kogan, M., Palen, L., & Anderson, K. M. (2015). Think local, retweet global: Retweeting by the geographically-vulnerable during Hurricane Sandy. In *Proceedings of the 18th ACM Conference on Computer Supported Cooperative Work & Social Computing* (pp. 981–993). New York, NY: ACM. https://dx.doi.org/10.1145/2675133.2675218

Kuppuswamy, V., & Bayus, B. L. (2018). A review of crowdfunding research and findings. In P. N. Golder & D. Mitra (Eds.), *Handbook of Research on New Product Development* (pp. 361–373). Northampton, MA: Edward Elgar Publishing.

Leavitt, A., & Clark, J. A. (2014). Upvoting Hurricane Sandy: Event-based news production processes on a social news site. In *Proceedings of the SIGCHI Conference on Human Factors in Computing Systems* (pp. 1495–1504). New York, NY: ACM Press. https://dx.doi.org/10.1145/2556288.2557140

Leavitt, A., & Robinson, J. J. (2017). The role of information visibility in network gatekeeping: Information aggregation on Reddit during crisis events. In *Proceedings of the 2017 ACM Conference on Computer Supported Cooperative Work and Social Computing* (pp. 1246–1261). New York, NY: ACM Press. https://dx.doi.org/10.1145/2998181.2998299

Li, J., Conathan, D., & Hughes, C. (2017). Rethinking emotional desensitization to violence: Methodological and theoretical insights from social media data. In *Proceedings of the 8th International Conference on Social Media & Society* (pp. 47:1–47:5). New York, NY: ACM. https://dx.doi.org/10.1145/3097286.3097333

Liu, S. B., Palen, L., & Giaccardi, E. (2012). Heritage matters in crisis informatics: How information and communication technology can support legacies of crisis events. In Christine Hagar (Ed.), *Crisis Information Management: Communication and Technologies* (pp. 65–86). Cambridge: Chandos Publishing.

Maddock, J., Starbird, K., Al-Hassani, H. J., Sandoval, D. E., Orand, M., & Mason, R. M. (2015). Characterizing online rumoring behavior using multi-dimensional signatures. In *Proceedings of the 18th ACM Conference on Computer Supported Cooperative Work & Social Computing* (pp. 228–241). New York, NY: ACM Press. https://dx.doi.org/10.1145/2675133.2675280

Mark, G., Al-Ani, B., & Semaan, B. (2009). Resilience through technology adoption: Merging the old and the new in Iraq. In *Proceedings of the 2009 Conference on Human Factors in Computing Systems (CHI 2009)* (pp. 689–698). New York, NY: ACM Press. https://dx.doi.org/10.1145/1518701.1518808

Mark, G., Bagdouri, M., Palen, L., Martin, J., Al-Ani, B., & Anderson, K. (2012). Blogs as a collective war diary. In *Proceedings of the 2012 Conference on Computer Supported Cooperative Work (CSCW 2012)* (pp. 37–46). New York, NY: ACM Press. https://dx.doi.org/10.1145/2145204.2145215

Mark, G., & Semaan, B. (2008). Resilience in collaboration: Technology as a resource for new patterns of action. In *Proceedings of the 2008 Conference on Computer Supported Cooperative Work (CSCW 2008)* (pp. 137–146). New York, NY: ACM Press. https://dx.doi.org/10.1145/1460563.1460585

Meier, P., & Brodock, K. (2008). *Crisis mapping Kenya's election violence: Comparing mainstream news, citizen journalism and Ushahidi* (Harvard Humanitarian Initiative). Boston, MA: Harvard University. Retrieved from: http://irevolution.wordpress.com/2008/10/23/mapping-kenyas-election-violence

Monroy-Hernández, A., boyd, d., Kiciman, E., De Choudhury, M., & Counts, S. (2013). The new war correspondents: The rise of civic media curation in urban warfare. In *Proceedings of the 2013 Conference on Computer Supported Cooperative Work* (pp. 1443–1452). New York, NY: ACM Press. https://dx.doi.org/10.1145/2441776.2441938

Morrow, N., Mock, N., Papendieck, A., & Kocmich, N. (2011). *Independent evaluation of the Ushahidi Haiti project.* Development Information Systems International. Retrieved from: www.alnap.org/pool/files/1282.pdf

Norheim-Hagtun, I., & Meier, P. (2010). Crowdsourcing for crisis mapping in Haiti. *Innovations: Technology, Governance, Globalization, 5,* 81–89. https://dx.doi.org/10.1162/INOV_a_00046

Okada, A., Ishida, Y., & Yamauchi, N. (2017). Effectiveness of social media in disaster fundraising: Mobilizing the public towards voluntary actions. *International Journal of Public Administration in the Digital Age, 4,* 49–68. https://dx.doi.org/10.4018/IJPADA.2017010104

Olteanu, A., Vieweg, S., & Castillo, C. (2015). What to expect when the unexpected happens: Social media communications across crises. In *Proceedings of the 18th ACM Conference on Computer Supported Cooperative Work & Social Computing* (pp. 994–1009). New York, NY: ACM Press. https://dx.doi.org/10.1145/2675133.2675242

Oyeyemi, S. O., Gabarron, E., & Wynn, R. (2014). Ebola, Twitter, and misinformation: A dangerous combination? *BMJ, 349,* g6178. https://dx.doi.org/10.1136/bmj.g6178

Palen, L., & Hughes, A. L. (2018). Social media in disaster communication. In H. Rodriguez, W. Donner, & J. E. Trainor (Eds.), *Handbook of disaster research* (pp. 497–518). Cham, Switzerland: Springer. https://dx.doi.org/10.1007/978-3-319-63254-4_24

Palen, L., Soden, R., Anderson, T. J., & Barrenechea, M. (2015). Success & scale in a data-producing organization: The socio-technical evolution of OpenStreetMap in response to humanitarian events. In *Proceedings of the 33rd Annual ACM Conference on Human Factors in Computing Systems* (pp. 4113–4122). New York, NY: ACM Press. https://dx.doi.org/10.1145/2702123.2702294

Palen, L., & Vieweg, S. (2008). The emergence of online widescale interaction in unexpected events. In *2008 ACM Proceedings of Computer Supported Cooperative Work Conference* (pp. 117–126). New York, NY: ACM Press. https://dx.doi.org/10.1145/1460563.1460583

Palen, L., Vieweg, S., & Anderson, K. M. (2011). Supporting "everyday analysts" in safety- and time-critical situations. *The Information Society, 27*(1), 52–62. https://dx.doi.org/10.1080/01972243.2011.534370

Palen, L., Vieweg, S., Liu, S. B., & Hughes, A. L. (2009). Crisis in a networked world. *Social Science Computing Review, 27,* 467–480.

Peary B. D. M., Shaw R., & Takeuchi Y. (2012). Utilization of social media in the East Japan earthquake and tsunami and its effectiveness. *Journal of Natural Disaster Science, 34,* 3–18. https://dx.doi.org/10.2328/jnds.34.3

Purohit, H., Hampton, A., Bhatt, S., Shalin, V. L., Sheth, A. P., & Flach, J. M. (2014). Identifying seekers and suppliers in social media communities to support crisis coordination. *Computer Supported Cooperative Work (CSCW), 23*(4–6), 513–545. https://dx.doi.org/10.1007/s10606-014-9209-y

Qu, Y., Wu, P. F., & Wang, X. (2009). Online community response to major disaster: A study of Tianya forum in the 2008 Sichuan earthquake. In *Proceedings of the 2009 Hawaii International Conference on System Sciences (HICSS 2009)* (pp. 1–11). Washington, DC: IEEE Computer Society. https://dx.doi.org/10.1109/HICSS. 2009.330

Reuter, C., Ludwig, T., Kaufhold, M.-A., & Pipek, V. (2015). XHELP: Design of a cross-platform social-media application to support volunteer moderators in disasters. In *Proceedings of the 33rd Annual ACM Conference on Human Factors in Computing Systems* (pp. 4093–4102). New York, NY: ACM Press. https://dx.doi. org/10.1145/2702123.2702171

Reuter, C., Marx, A., & Pipek, V. (2012). Crisis management 2.0: Towards a systematization of social software use in crisis situations. *International Journal of Information Systems for Crisis Response and Management, 4*, 1–16. https://dx.doi.org/10.4018/ jiscrm.2012010101

Schmidt, A., Wolbers, J., Ferguson, J., & Boersma, K. (2017). Are you Ready-2Help? Conceptualizing the management of online and onsite volunteer convergence. *Journal of Contingencies and Crisis Management, 26*, 338–339. https://dx.doi. org/10.1111/1468-5973.12200

Schulten, K. (2015, October 2). Skills and strategies: Fake news vs. real news: Determining the reliability of sources. *New York Times.* Retrieved from: https://learning. blogs.nytimes.com/2015/10/02/skills-and-strategies-fake-news-vs-real-news-determining-the-reliability-of-sources/

Scott, R. (2017, December 6). Charity scams put the "disaster" in disaster relief. *Huffington Post.* Retrieved from: www.huffingtonpost.com/ryan-scott/charity-scams-put-the-dis_b_8248574.html

Semaan, B., & Mark, G. (2012). "Facebooking" towards crisis recovery and beyond: Disruption as an opportunity. In *Proceedings of the ACM 2012 conference on Computer Supported Cooperative Work* (pp. 27–36). New York, NY: ACM Press. https://dx.doi. org/10.1145/2145204.2145214

Shaw, F., Burgess, J., Crawford, K., & Bruns, A. (2013). Sharing news, making sense, saying thanks: Patterns of talk on Twitter during the Queensland Floods. *Australian Journal of Communication, 40*, 23–40. https://core.ac.uk/download/pdf/30676051. pdf

Soden, R., & Palen, L. (2014). From crowdsourced mapping to community mapping: The post-earthquake work of OpenStreetMap Haiti. In *Proceedings of the 11th International Conferences on the Design of Cooperative Systems (COOP 2014)* (pp. 311–326). Nice, France: Springer. https://dx.doi.org/10.1007/978-3-319-06498-7_19

Soden, R., & Palen, L. (2016). Infrastructure in the wild: What mapping in post-earthquake Nepal reveals about infrastructural emergence. In *Proceedings of the 2016 CHI Conference on Human Factors in Computing Systems* (pp. 2796–2807). New York, NY: ACM Press. https://dx.doi.org/10.1145/2858036.2858545

St. Denis, L. A., Hughes, A. L., & Palen, L. (2012). Trial by fire: The deployment of trusted digital volunteers in the 2011 Shadow Lake Fire. In *Proceedings of the Information Systems for Crisis Response and Management Conference (ISCRAM 2012).* Vancouver, BC. Retrieved from http://idl.iscram.org/files/stdenis/2012/207_St. Denis_etal2012.pdf

St. Denis, L. A., Palen, L., & Anderson, K. M. (2014). Mastering social media: An analysis of Jefferson County's communications during the 2013 Colorado Floods. In *Proceedings of the Information Systems for Crisis Response and Management Conference (ISCRAM 20014)*. Retrieved from http://iscram2014.org/sites/default/files/misc/proceedings/p93.pdf

Starbird, K. (2017). Examining the alternative media ecosystem through the production of alternative narratives of mass shooting events on Twitter. In *Proceedings of the Tenth International AAAI Conference on Web and Social Media (ICWSM 2017)*. Montreal, Canada. https://pdfs.semanticscholar.org/3e90/59740ca3c54b213b3f70e91a503afaa271af.pdf?_ga=2.238987904.1888840902.1539903415-1843962977.1539903415

Starbird, K., & Palen, L. (2011). "Voluntweeters": Self-organizing by digital volunteers in times of crisis. In *Proceedings of the 2011 Conference on Human Factors in Computing Systems (CHI 2011)* (pp. 1071–1080). New York, NY: ACM Press.

Starbird, K., & Palen, L. (2012). (How) Will the revolution be retweeted? Information propagation in the 2011 Egyptian uprising. In *Proceedings of the 2012 Conference on Computer Supported Cooperative Work (CSCW 2012)* (pp. 7–16). New York, NY: ACM Press. https://dx.doi.org/10.1145/2145204.2145212

Starbird, K., & Palen, L. (2013). Working & sustaining the virtual "disaster desk." In *Proceedings of the 2013 Conference on Computer Supported Cooperative Work (CSCW 2013)* (pp. 491–502). New York, NY: ACM Press. https://dx.doi.org/10.1145/2441776.2441832

Stephens, K. K., & Malone, P. C. (2009). If the organizations won't give us information... The use of multiple new media for crisis technical translation and dialogue. *Journal of Public Relations Research, 21*, 229–239. https://dx.doi.org/10.1080/10627260802557605

Thomson, R., Ito, N., Suda, H., Lin, F., Liu, Y., Hayasaka, R., Isochi, R., & Wang, Z. (2012). Trusting tweets: The Fukushima disaster and information source credibility on Twitter. In *Proceedings of the Information Systems for Crisis Response and Management Conference (ISCRAM 2012)*. Vancouver, BC.

Tonkin, E., Pfeiffer, H. D., & Tourte, G. (2012). Twitter, information sharing and the London riots? *Bulletin of the American Society for Information Science and Technology, 38*, 49–57. https://dx.doi.org/10.1002/bult.2012.1720380212

Vieweg, S., Palen, L., Liu, S. B., Hughes, A. L., & Sutton, J. (2008). Collective intelligence in disaster: Examination of the phenomenon in the aftermath of the 2007 Virginia Tech Shooting. In *Proceedings of the Information Systems for Crisis Response and Management Conference (ISCRAM 2008)*. Washington, DC. Retrieved from: http://idl.iscram.org/files/vieweg/2008/1051_Vieweg_etal2008.pdf

White, J. I., & Palen, L. (2015). Expertise in the wired wild west. In *Proceedings of the 18th ACM Conference on Computer Supported Cooperative Work & Social Computing* (pp. 662–675). New York, NY: ACM Press. https://dx.doi.org/10.1145/2675133.2675167

10

DORMANT DISASTER ORGANIZING AND THE ROLE OF SOCIAL MEDIA

Chih-Hui Lai

> A very common response to environmental stress is for organisms to enter a reversible state of reduced metabolic activity, or dormancy. By doing so, these organisms can drastically lower their energetic expenditures and evade unfavorable conditions that would otherwise reduce the fitness of the population. Dormancy is not a cost-free strategy, however. Organisms must invest resources into resting structures and the machinery that is needed for transitioning into and out of a dormant state.
>
> (Lennon & Jones, 2011, p. 119)

The above excerpt is the definition of dormancy from the field of biology. Animals go into a dormant diapause as a mechanism of adaptation in response to the scarcity of the resources available in the environment (Levins, 1968). Dormancy is seen as an evolved ability of microbial species to adapt during periods of environmental stress and as "a reservoir of dormant individuals that can potentially be resuscitated in the future under different environmental conditions" (Lennon & Jones, 2011, p. 119). The machinery to bring an organism into and out of a dormant state may include relationships and the technologies that support the maintenance of the relationships.

Human organized behavior is triggered by a common focus of activity (Feld, 1981). For non-emergency tasks, it is not uncommon that people form a group for the task and once the task is accomplished, the group becomes dormant. For example, a group of college students work on a class project and create a group page on Facebook. After the project is completed, the group page is still up, albeit with idle activity.

In events of mass convergence or emergencies, different forms of human organized behavior are observed, ranging from formal non-profit and public organizations, to ad hoc, self-organized, citizen-based groups. When the exigencies are over, these organized behaviors seemingly retreat out of sight. For example, a group of volunteers is temporarily mobilized for relief operations for an earthquake through a Twitter page and no activities are updated two months after the disaster. One of the possibilities about the inactive group page is that the group identity as a response group is shadowed, but members keep connected to each other through face-to-face encounters or other electronic modes of communication as a loosely connected group. The maintenance of these connections can be used as the foundation of reactivating the group for a new disaster event. In sum, dormant disaster organizing refers *to the process of citizen-based response groups maintaining latent identity and relationships after the task of disaster response is accomplished, and reactivating when a new disaster event requires the assemblage.*

Dormant disaster organizing is not a new or unique phenomenon; however, due to the use of social media the observation of dormant disaster organizing has become more evident. While research has delved into various hidden and invisible forms of organizations, such as terrorist organizations or secret societies (e.g., Jones, 2006; Mishal & Rosenthal, 2005), little attention has been given to the phenomenon of dormant organizing in general and in the context of disaster response in particular. With the growing importance of social media use in citizen-based disaster response groups (Lai, 2017; Starbird & Palen, 2011), there is a need to draw on existing theoretical frameworks to explain and understand the process of how citizen-based response groups become dormant after the disaster, and reactivate when a new disaster happens. To that end, this chapter introduces the concept of dormant disaster organizing.

This chapter aims to answer the following questions by untangling the relational and technological systems of dormant disaster organizing:

1. What constitutes dormant organizing?
2. Why does dormant disaster organizing happen?
3. How is dormant disaster organizing maintained?

The ultimate purpose is to offer an informed societal understanding about the potential of preparing organized efforts for future disaster response. This endeavor presents an important extension to existing research on organizational communication in general and fluid forms of organized social behavior in particular.

Essentially, increasing the knowledge about how dormant disaster organizing is practiced helps citizen-based response groups maintain their organizational status between disaster events. This has implications for disaster preparedness and response because physical and volunteer communities can learn how to build capacities between disasters to more robustly adapt to future exigencies. In other words, this conceptualization offers the idea that groups deactivating between exigencies may be, in fact, *alternative forms of organizing* shaping societal well-being in the long run (Parker, Cheney, Fournier, & Land, 2014; Scott, 2013).

In this chapter, a review of the applicability of existing theories on organized behavior to explain dormant disaster organizing is provided. This review is followed by the presentation of the conceptualization of dormant disaster organizing building on the theories of ecology and evolution, coupled with relevant middle-range theories, including latent tie theory (Haythornthwaite, 2002), activity focus theory (Feld, 1981), and information and communication public goods theory (Fulk, Flanagin, Kalman, Monge, & Ryan, 1996). The chapter concludes by presenting practical implications and discussion questions concerning dormant disaster organizing.

Extant Research on Organized Behavior

Before explicating the process of dormant disaster organizing, it is important to point out that theorizing about dormant disaster organizing sits at the intersection of three areas of research. First, a rich line of theoretical frameworks in the area of organizational communication delves into the process of inception, maintenance, and demise of organized behavior (Monge, Heiss, & Margolin, 2008). Nonetheless, despite the recognition of flexible boundaries and fluid forms of organizing (Cooper & Shumate, 2012; Walker & Stohl, 2012), one of the assumptions of these frameworks is that organizations exist because of members' consistent participation in organizational activity. Yet it is unclear whether the presence of all or a few organizational members is sufficient to ensure the existence of the organization as it evolves.

Second, another line of research on self-organized social behavior in non-routine contexts, such as social movements and emergency response, identifies the mechanisms of emergent collectives (Saunders & Kreps, 1987), and a few focus on tracing the social roots of these collectives (Chadwick, 2007). Yet the theoretical focus of the research is on how these collectives come into being, triggered by environmental events, particularly through the use of information and communication technologies (ICTs) (e.g., mobile phones, social media). The presumed value of these collectives resides in their existence in relation to the ongoing event. Less attention is paid to the mechanisms of whether and how these collectives are maintained after the event is concluded or between events.

Third, research on computer-mediated communication (CMC) and virtual communities investigates the combined influence of technological features and reduced personal cues on the enhanced or intensified social interaction online, be it positive or negative (Baym, 2007; Rheingold, 2000; Walther, 1996). Yet the assumption is that membership is limited to a constant group identity or purpose that brings members together. The identity or the purpose itself hardly changes over time.

What Is Dormant Organizing?

Integrating and filling the gaps of the aforementioned three lines of research, dormant disaster organizing is considered as a process of loosely connected groups of individuals coming together because of dedication to a disaster event, and after the event, somehow sustaining the relational and technological foundation of the group, which is subject to regrouping when necessary. This definition reflects the following characteristics:

1. ICTs play a key role in facilitating the emergence of citizen-based response groups' organizing when the disaster happens as well as maintaining the group after the disaster.
2. Groups' disaster organizing becomes dormant because collective activities have decreased or taken a different turn, which renders the original goal of the group less salient.
3. Dormant disaster organizing embodies the ability of "just in time" reactivation, meaning that the group can be mobilized and reactivate when a new triggering disaster event happens.

To understand dormant disaster organizing, it is necessary to unpack the evolutionary process of how disaster organizing comes into being in the first place.

Ecological and Evolutionary Theories of Organized Behavior

Rooted in biology, ecological and evolutionary theories of organizations consider that the ways an organization is connected to its environment has consequences on the functioning and performance of the organization. In Homans' (1950) early remarks on social organizations, he pointed out that when studying a social system, "what is the nature of the group's environment?" is the first question the researcher needs to ask. Moreover, it is necessary to consider the adaptive processes in characterizing human groups. In sum, the focus is on the dual process of how a social organization is affected by the surrounding environment and how it adapts itself to the environment by obtaining resources. These two inquiries lay the foundation of the ecological and evolutionary perspective, respectively; that is, *selection* and *adaptation*, which have become integrated in the research that follows (Baum & Shipilov, 2006). An ecological perspective focuses on the selection mechanism of certain organizational characteristics that fit the environmental demands (e.g., Hannan & Freeman, 1977; Kaufman, 1975), while the evolutionary counterpart highlights the active and adaptive processes enacted by organizations as they evolve (Burgelman, 1991; Meyer, 1994). An integrated approach of the two perspectives explains how organizational founding, failure, and change is a result of the influence of the social environment, as well as the adaptive interactions among organizations and/or populations of organizations (Baum & Shipilov, 2006). It examines the process of how organizations pursue the goal of fitness by means of interacting with other members in their communities and populations as well as interacting with their environments (Campbell, 1965; Hannan & Freeman, 1977, see Monge et al., 2008 for a review of ecological and evolutionary theories in organizational research).

Instead of theorizing environments as a priori, Weick's (1979) ecological model of organizing contends that the environment exists through organizational actors' retrospective interpretations of actions/retrospective attentional processes. Hence, actors adapt to the environment that they create. An organization can be defined in terms of processes of organizing, which are directed toward information processing, and in particular, removal of equivocality in the information environment enacted by actors of the organization (Weick, 1969). Moreover, the processes of organizing rely on interlocked behavior (Weick, 1969). That is, individual behavior is contingent on the behavior of others. Such interdependent and interlocked behavior are critical in resolving equivocality, which requires actors to interlock sets of their behavior in order to produce certainty. "Individuals come together because each wants to perform some act and needs the other person to do certain things in order to make performance possible" (Weick, 1969, p. 91).

According to Weick, a group or an organization is formed after actors show a convergence of shared ideas and interest, which then activate the collective structure in the form of a repetitive cycle of interlocked behaviors. These interlocked behaviors are then embedded in enactment (variation), selection, and retention (VSR) processes. Variations consist of a repertoire of responses and the selection process involves the enactment of interlocked cycles to select meaning that is imposed on these equivocal inputs interpreted by humans (Weick, 1979). The selected meaning is stored in the retention process in the form of a cause map, which will determine how the remembered items will be used in the future activities. In short, VSR processes are key to the evolution of groups or organizations. For example, a group experiments with a set of options in terms of using social media to coordinate group activities (variation), give meaning to those options based on their usefulness (selection), and retain the optimal ones for future coordination (retention).

Why Dormant Organizing: From an Ecological and Evolutionary View

In Weick's (1979) view, evolution represents raw change not necessarily in the direction of increasing orderliness. Retained contents can also be internally inconsistent. Hence, internal reorganization is necessary, which explains why dormant disaster organizing is enacted. When a disaster happens, individuals show a convergence of shared ideas and interests, which then activate the collective structure in the form of a repetitive cycle of interlocked behaviors. Citizen-based response groups are formed, followed by members' enactment of the disaster environment, selecting the strategies of handling the environment, and retaining the strategies that work. In most cases, disaster response groups' Facebook or Twitter pages may serve to store the memory of the VSR processes of organizing. With the newly enacted environment after the disaster, however, those response groups need to select new meaning for dealing with the post-disaster situation, and contemplate whether the previously retained strategies become inconsistent or still applicable to the group. These processes are engaged in the form of interlocked behavior among group members and thus a new cycle of organizing is practiced.

At the broader level, dormant disaster organizing also involves the relationships with other entities in the environment. In ecological and evolutionary theories, the concept of quantum speciation refers to the process where new species spawn from old ones under fortuitous conditions; for example, mutant individuals are isolated by geographic barriers and form different species (Astley, 1985; Grant, 1963). In other words, ecological opportunities must present themselves for mutual populations to form and occupy a space in the environment (Stanley, 1981). Disasters present such ecological opportunities for the emergence of citizen-based response groups. During the disaster, citizen-based response groups are visible to the extent that they occupy a space in the environment with other pre-existing nonprofit and public organizations. After the disaster, if response groups engage in dormant organizing (e.g., group members maintaining ties with each other, or engaging in other activity while keeping the group intact), their space can be reserved, and their relationships with other pre-existing organizations are maintained, which helps their mobilization in the future.

Why Dormant Organizing: From an Activity Focus View

Integral to the ecological and evolutionary theories is the network relationships between individual members, or between groups/organizations (Monge et al., 2008). According to Feld's activity focus theory, individuals are likely to develop ties with each other by participating in joint activities, namely, foci, or "social, psychological, legal, or physical entit[ies] around which joint activities are organized (e.g., workplaces, voluntary organizations, hangouts, families, etc.)" (Feld, 1981, p. 1016). Specifically, if under pressures of time and emotion, individuals are more likely to combine their interactions by exploring and developing new foci to bring more of other network contacts together and engage in joint activities (Feld, 1981). A particular set of focused activities motivate people to develop ties, which may further facilitate individuals' participation in other types of focused activities or develop new ones. Similarly, individuals are likely to join an activity engaged by other people with whom they have relationships. Co-participation in focused activities thus allows for the observation of the reciprocal process of communicative links being created and reproduced (McPhee & Corman, 1995).

Corman and Scott (1994) proposed that perceived network relationships are connected to observable communication through activities. Latent networks of communication relationships can be triggered by activity and become observable communication. They defined a network as "a set of rules and resources actors draw upon in accomplishing communicative interaction, which interacts in critical ways with features of the collective context to determine who is likely to talk with whom in a given instance" (Corman & Scott, 1994, p. 172). When communication is instantiated, those structural links are reproduced (Corman & Scott, 1994). Accordingly, both network links and foci are necessary to trigger communicative actions, because when the focus is triggered (e.g., disaster), it organizes activity conducive to communication. That also means that if the focus is not active, it ceases to organize activity and communication becomes dormant.

Based on activity focus theory, citizen-based response groups' disaster organizing becomes dormant because the original focus is not active. Nonetheless, the network links established during the disaster can motivate individuals to participate in other focused activities. Those links can also become latent, waiting to be triggered by a new focus (e.g., another new disaster). In this regard, research on latent organizations and latent ties can shed light on the process of how dormant disaster organizing moves from dormancy to reactivation. Latent organizations refer to the "groupings of individuals and teams of individuals that persist through time and are periodically drawn together for recurrent projects" (Starkey, Barnatt, & Tempest, 2000, p. 299). Latent organizations arguably allow for coordination and collaboration beyond hierarchy and vertically integrated modes in more flexible network forms of organization, especially in risky and uncertain environments (Ebbers & Wijnberg, 2009). Over time, members lay both relational and structural foundations for deactivation even when they no longer actively work together toward the same goal; but they then regroup whenever the opportunity arises.

The decentralized, flexible, and loosely connected characteristics of latent organizations are similar to those of dormant disaster organizing. But unlike Starkey et al.'s (2000) conceptualization, which posits that latent organizations consist of the same members who disband and reactivate when the project demands, citizen-based response groups may be open to a wide scope of membership from within and outside the affected areas every time the group reactivates. That also means that after each time of reactivation, members who have latent ties with each other may change.

Latent ties refer to those technically feasible yet not socially activated social ties (Haythornthwaite, 2002). For example, an individual develops latent ties with other group members by joining this group's Facebook page. Only after this person initiates certain social interactions with other members can these latent ties transform into active ties. Unlike entirely new potential ties, latent ties represent the social foundation which can be used for new situations. Mariotti and Delbridge's (2012) study on the European motorsport industry showed that potential ties (embryonic relationships where interaction has not yet occurred) and latent ties (established relationships but not are currently active) are part of the strategic actions taken by firms to cope with network redundancy and overload. They argue that this contributes to a dynamic view of network evolution because "networks thus evolve through actors looking at new opportunities (developing potential ties) and/or by suspending exchange with others that are redundant (maintaining latent ties till circumstances arise)" (Mariotti & Delbridge, 2012, p. 512). After a period of dormancy, latent ties may provide novel and non-redundant information to the collective (Mariotti & Delbridge, 2012). As a result, through members' latent ties and the overall group latent identity, dormant disaster organizing develops the capacity for future reactivation when needed.

How to Maintain Dormant Organizing: A Collective Action View

After illustrating the reasons why dormant disaster organizing happens, next I explain how dormant organizing is maintained. When citizen-based response groups engage in disaster relief actions, the digital artifacts created, such as the posts on Twitter and Facebook pages, and the social relationships and practices together constitute resources that allow groups to share and exchange information, provide and receive emotional support, and coordinate activity (Resnick, 2001). After the disaster, groups are maintained through the low maintenance of social ties and communication infrastructures. As Resnick (2001) argued, the technological mechanisms of allowing a group to go dormant and to be able to reassemble when it is needed should be considered simultaneously.

What exactly does social media afford the ability of maintaining and reactivating social ties and groups when necessary? The theory of information and communication public goods can provide important insights. Fulk et al (1996) proposed that two forms of public goods, communal (information sharing) and connective (direct connections) goods, can be generated through use of organizational information repositories (Fulk, Heino, Flanagin, Monge, & Bar, 2004; van den Hooff, 2004). Connectivity contains physical and social connectivity. The former refers to the level of technological infrastructure and access, and the latter represents the degree of social relationships (Fulk et al., 1996). Physical connectivity is necessary for social connectivity to materialize. For example, members of a citizen-based response group will not be able to connect to each other without the intact telecommunications infrastructures that make phone calls or social media uses possible.

Applying concepts from the theory of information and communication public goods to dormant disaster organizing, it is expected that after the disaster, citizen-based response groups may keep the group's social media page alive, which allows for the maintenance of communality and connectivity among group members. Specifically, the communal information contains the detailed organizing process during the disaster whereas connectivity is facilitated as members still publicize their affiliation with the group and thus direct communication is possible. While this may not always be intentional (e.g., a group may simply fail to close the page, rather than consciously choosing to leave it up), it still affords the potential for communality and connectivity to the group members.

According to information and communication public goods theory, lacking updated information from individual contributors in a distributed information system may render the communal good less attractive and less valuable to potential contributors (Monge et al., 1998). Yet, in the case of dormant disaster organizing, the quality of communal information may not be measured in terms of timeliness, and instead should be measured by completeness and replicability. Despite the fact that the communal information about what the group was doing during the disaster is outdated, it provides a place for the group itself as well as interested others to learn about the procedures of mobilization, which can be used for similar actions next time. Moreover, the information on groups' Facebook or Twitter pages documenting how the group provided support to the affected communities can serve as a way to retain the memory of the latent identities of the group (Pratt & Foreman, 2000). Therefore, these practices of storing latent identities create a structural artifact of the group, making the organizing process more transparent and accessible for future disaster response (Meier, 2014).

Transparency and durability afforded through social media allows for existing members to record the organizing process and maintain the latent relational structure. For new and potential members, this transparency and accessibility provides an opportunity to learn about previously enacted coordinating mechanisms during the disaster. Consequently, even when the response groups are relatively unknown or invisible after the disaster, their coordination can be successfully sustained and immediately acted on by members when a new disaster event occurs. Furthermore, via the same platform or other private channels, a smaller group of members may engage in closer communication and create an information pool on topics other than disaster response.

According to my observations of recent disasters, for example, after Hurricane Sandy, citizen-based response groups maintained their group page on Facebook. A few months after Sandy, a tornado happened and the groups quickly reactivated their connections and either created a new group page on Facebook or recycled the original Sandy page to mobilize support for the survivors (Lai, 2017). Hurricane Sandy, which occurred on October 29, 2012, devastated much of the northeast of the USA, with particularly destructive damage occurring in New York and New Jersey. With the estimated cost of the damage as $50 billion, it was considered to be the second-costliest hurricane on record in the USA after Hurricane Katrina in 2005 (Blake, Kimberlain, Berg, Cangialosi, & Beven, 2013).

In fact, as a result of persistent contact and pervasive awareness afforded by social media, social relationships and the context where those relationships are formed become more enduring and people live in a society characterized by pervasive surveillance combined with intensive subtle ways of information exchange and information about social ties (Hampton, 2016). Hampton argues that because of social media, social ties are continued over time regardless of life changes. Extending this conceptualization to the group level, it can be argued that the affordance of preservation of social ties enables dormant disaster organizing, because citizen-based response groups mobilized for a particular disaster event are preserved and the ties among members can be maintained on an interpersonal or group level. Certainly, those ties between members can become latent, but the pervasive awareness of each other's statuses can be an informal way of collecting information about each other (Hampton, 2016). Hence, when a new disaster event happens, the ties can be quickly reactivated. After the disaster, the ties between members can also be enhanced through other focused activities, which results in higher rates of bridging between and within foci (Hampton, 2016).

A Typology of Dormant Disaster Organizing

In sum, dormant disaster organizing is enabled in part due to the preservation and persistence of social ties afforded by social media (Treem & Leonardi, 2013), which store the memory of the previous organizing process, including the selection process of interpreting the equivocal inputs and the retention process of storing the enacted environments and cause maps. Before the advent of social media, these processes of organizing were mostly likely kept in the minds of individual members or group documents, if possible. Yet they were not easily accessible by the members who were involved or who wanted to be involved when a new disaster event occurs, or used for repeated occasions. With social media, the content may be open to the public, instead of only to a few members. The retained content in the form of existing group pages serve to maintain a sense of stability so when enacting future disaster environments, individuals are able to find the convergence of interests and ideas quickly. Instead of totally starting from scratch (nothing is retrieved from the group memory) or being constrained by the previous enactments (retrieved everything from the memory without considering the task at hand), dormant disaster organizing offers a mechanism of balancing stability and flexibility when triggered by a new disaster event. While acting following the previous retained strategies when a new disaster happens, response groups also gain new perspectives built through members' latent (or active) ties maintained between disasters.

Building on Feld's activity focus theory and Weick's ecological model of organizing, I propose a typology of dormant disaster organizing. Figure 10.1 explains four types of dormant disaster organizing based on the two dimensions—*adaptive strategies of foci (preserved, morphed) and engagement of interlocked behavior (latent, ongoing)*. The first dimension describes the degree to which a group upholds the focused activity that brings the group together, which can be considered as either being maintained (preserved) or being transformed (morphed). The second dimension captures the extent to which the members are involved in explicit (ongoing) or implicit (latent) collective behavior that impacts the organizing of the group.

Applying activity focus theory and the ecological model of organizing, I argue that citizen-based response groups are assembled because of the joint activities triggered by the disaster and social connections built among members. Additionally, members interlock their behaviors to produce certainty in response to the equivocality that arises as the disaster happens. After the disaster, groups' activity focus is preserved yet becomes dormant. The links among members become latent as well, waiting to be triggered by a new focus (e.g., another new disaster). On the other hand, the links built during the disaster may motivate members to explore other types of focused activity. In this way, groups' foci may deviate from the original one that brought the group together in the first place, but it allows for the collective interlocked behavior to continue to be performed in a new context. Accordingly, these two dimensions—activity focus and interlocked behavior—account for the strategies of how citizen-based response groups adapt themselves to the new equivocality as a result of the termination of the disaster event.

Following the two dimensions, four types of dormant disaster organizing are identified. Type 1 groups keep the social media group page but make no update after a few months following the disaster event. Member connections are discontinued as well. Similar to Type 1, Type 3 groups preserve the social media group page, but members keep in touch with each other outside the group on a dyadic basis and engage in other focused activities relevant to their personal interests. Both Type 2 and Type 4 groups keep the social media group page and ongoing group activities, but Type 2 groups engage in similar focused activities related to the disaster event. For example, a disaster response group will post messages elated to other disaster actions, or emergency relief. With the same group name and social media group page, Type 4 groups engage in different collective activities. For example, a response group posts updates about a social movement or mobilize support for a social cause.

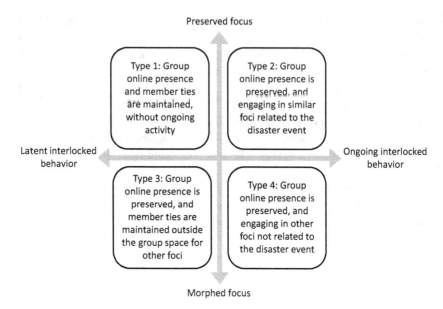

FIGURE 10.1 Types of dormant disaster organizing

Types of Dormant Disaster Organizing Exhibited by Each Group

Note that these four types of groups exhibit different levels of dormant disaster organizing. Type 2 has the highest level of capacity for reactivating the group for a new disaster event, followed by Type 3 and Type 4. For Type 2 groups, members continue building their repertoires of strategies such as the ways to mobilize resources and communicate within and with other entities; when they have a chance to regroup, they can recombine their practices of repertoires and carry out the collective action effectively and efficiently. Compared to Type 2 groups, in Type 3 groups, members' ongoing dyadic interaction after the disaster will be less effective in facilitating collective action when needed. This is because the interaction typically takes place between members only and is less ready to be configured to produce interlocked behavior on a collective level. In the meantime, despite the ongoing interlocked behavior among group members, because of the change of activity focus, Type 4 groups may have experienced change in certain aspects of organizing, ranging from memberships, norms, expectations, and goals, which makes reactivation for a new disaster challenging. Even with the preservation of group focus, the lack of ongoing interlocked behavior among members renders Type 1 groups least effective in mobilization in response to a new disaster.

The proposed model emphasizes the importance of paying attention to not only the existence of dormant disaster organizing but also the mechanisms underlying dormant disaster organizing. Building on the four dimensions, four types of dormant disaster organizing are identified, which can be further linked to varied capacities of reactivating when a new disaster event happens. Note that dormant disaster organizing might not necessarily be a deliberate choice. Yet dormant disaster organizing is the means that a group of people uses to make sense of the environment that is different from the one that brings them together. Although it sounds like an oxymoron, it represents the complex and latent nature of human behaviors and relationships.

Practical Implications

This chapter on dormant disaster organizing has practical implications for disaster management as the knowledge of whether, how, and when relief support can be mobilized and channeled toward necessary response tasks can help disaster preparedness and build community resilience. Community resilience refers to the capacities of a community to prepare, respond, and recover after emergencies (Norris, Stevens, Pfefferbaum, Wyche, & Pfefferbaum, 2008; Pfefferbaum et al., 2013). These capacities entail the extent to which social connections and organizational networks can help individuals and communities find support at times of crises (Sherrieb, Norris, & Galea, 2010). In other words, facilitating and sustaining citizen-based response groups' dormant disaster organizing—for example, through detailed documentation of the response efforts during the disaster and maintaining member connections after the disaster through social media—helps foster response groups' resilience capacity-building, which in turn enhances local communities' resilience. Moreover, these mechanisms of facilitating dormant disaster organizing strengthen transparency and accessibility of citizen-based response groups' past organizational practices, which are helpful for people who are interested in volunteering as well as public and nonprofit organizations who may collaborate with response groups for future relief actions.

Although the focus of this chapter is on social media, dormant disaster organizing may take non-mediated forms. For example, member ties may be maintained through in-person neighborhood meetings and the receipt and exchange of information on a local bulletin board. Indeed, when citizen-based response groups are activated, they often consist of members from inside and outside the affected community (Wenger & James, 1994). For members from outside the local community, having connections with local ties is necessary in order to know the effective way to deliver resources to the most needed (Lai, Tao, & Cheng, 2017). Yet it is rare that disasters happen to the same community within a short period of time, which means that memberships of the same response groups may change after every disaster. As such, locality-based, non-mediated dormant disaster organizing may serve less useful functions than those enabled online. As such, from the perspective of disaster response, dormant disaster organizing enabled through social media has more flexibility and utility for future disaster response than do non-mediated forms.

This chapter addresses the questions of what constitutes dormant disaster organizing, why dormant disaster organizing happens, and how dormant disaster organizing is maintained. Based on latent tie theory (Haythornthwaite, 2002), activity focus theory (Feld, 1981), and information and communication public goods theory (Fulk et al., 1996), a typology of dormant disaster organizing is proposed based on two dimensions: activity focus and interlocked behavior. This creates four types of dormant disaster organizing that represent the combination of preserved/morphed focus and ongoing/latent interlocked behavior. I also illustrate how the four types of dormant disaster organizing can be used to observe groups' varied capacities of reactivating when a new disaster event happens.

Discussion Questions

1. In this chapter, a typology of four types of dormant disaster organizing is proposed to account for the various ways dormant groups might reactivate in response to a new disaster event. Several questions merit further discussion. For example, compared to citizen-based response groups' dormant disaster organizing, how would public and nonprofit organizations experience similar or different processes of dormant organizing between disasters? How does that in turn influence citizen-based response groups' dormant disaster organizing?

2. Social media use in disaster self-organizing contexts is, in some ways, similar to enterprise social media use because it allows for the creation of shared affordances. That is, group members converge on the use of technological features in order to harness the joint affordances of the technology and coordinate group action (Ellison, Gibbs, & Weber, 2015; Leonardi, 2013). To what extent does social media enable and constrain dormant organizing between disasters?

3. A disaster can inarguably take different forms, including natural (e.g., earthquakes), public health crisis (e.g., Ebola), and human-induced (e.g., chemical incidents). These disasters embody different temporal characteristics. For example, the entire process of emergency response to an epidemic may take several months, while rescuing people from debris caused by a hurricane might take several days. How would dormant disaster organizing be affected by different forms of disasters? How might this impact the connections between group members and other stakeholders?

References

Astley, W. G. (1985). The two ecologies: Population and community perspectives on organizational evolution. *Administrative Science Quarterly, 30*(2), 224–241. www.jstor.org/stable/2393106

Baum, J. A. C., & Shipilov, A. V. (2006). Ecological approaches to organizations. In S. R. Clegg, C. Hardy, W. Nord, & T. Lawrence (Eds.), *Sage handbook for organization studies* (pp. 55–110). London: Sage.

Baym, N. K. (2007). The new shape of online community: The example of Swedish independent music fandom. *First Monday, 12*(8). http://journals.uic.edu/ojs/index. php/fm/article/view/1978/1853

Blake, E. S., Kimberlain, T. B., Berg, R. J., Cangialosi, J. P., & Beven, J. L. II (2013). Tropical cyclone report, Hurricane Sandy: October 22–29, 2012. *National Hurricane Center.* Retrieved from: www.nhc.noaa.gov/data/tcr/AL182012_Sandy.pdf

Burgelman, R. A. (1991). Intraorganizational ecology of strategy making and organizational adaptation: Theory and field research. *Organization Science, 2*(3), 239–262. https://doi.org/10.1287/orsc.2.3.239

Campbell, D. T. (1965). Variation and selective retention in socio-cultural evolution. In H. R. Barringer, G. I. Blanksten, & R. W. Mack (Eds.), *Social change in developing areas: A reinterpretation of evolutionary theory* (pp. 19–48). Cambridge, MA: Schenkman.

Chadwick, A. (2007). Digital network repertoires and organizational hybridity. *Political Communication, 24*(3), 283–301. http://doi.org/10.1080/10584600701471666

Cooper, K. R., & Shumate, M. (2012). Interorganizational collaboration explored through the bona fide network perspective. *Management Communication Quarterly, 26*(4), 623–654. http://doi.org/10.1177/0893318912462014

Corman, S. R., & Scott, C. R. (1994). Perceived networks, activity foci, and observable communication in social collectives. *Communication Theory, 4*, 171–190. http://dx.doi.org/10.1111/j.1468-2885.1994.tb00089.x

Ebbers, J. J., & Wijnberg, N. M. (2009). Latent organizations in the film industry: Contracts, rewards and resources. *Human Relations, 62*(7), 987–1009. http://dx.doi.org/10.1177/0018726709335544

Ellison, N. B., Gibbs, J. L., & Weber, M. S. (2015). The use of enterprise social network sites for knowledge sharing in distributed organizations: The role of organizational affordances. *American Behavioral Scientist, 59*(1), 103–123. http://dx.doi.org/10.1177/0002764214540510

Feld, S. (1981). The focused organization of organizational ties. *American Journal of Sociology, 86*, 1015–1035. www.jstor.org/stable/2778746

Fulk, J., Flanagin, A. J., Kalman, M. E., Monge, P. R., & Ryan, T. (1996). Connective and communal public goods in interactive communication systems. *Communication Theory, 6*, 60–87. http://dx.doi.org/10.1111/j.1468-2885.1996.tb00120.x

Fulk, J., Heino, R., Flanagin, A. J., Monge, P. R., & Bar, F. (2004). A test of the individual action model for organizational information commons. *Organization Science, 15*, 569–585. http://dx.doi.org/10.1287/orsc.1040.0081

Grant, V. (1963). *The origin of adaptations.* New York: Columbia University Press.

Haythornthwaite, C. (2002). Strong, weak and latent ties and the impact of new media. *The Information Society, 18*(5), 385–401. http://dx.doi.org/10.1080/019722 40290108195

Hampton, K. N. (2016). Persistent and pervasive community: New communication technologies and the future of community. *American Behavioral Scientist, 60*(1), 101–124. http://doi.org/10.1177/0002764215601714

Hannan, M. T., & Freeman, J. (1977). The population ecology of organizations. *American Journal of Sociology, 82*(5), 929–964. www.journals.uchicago.edu/doi/abs/10.1086/226424

Homans, G. C. (1950). *The human group.* New York: Harcourt, Brace.

Jones, C. (2006). Al-Qaeda's innovative improvisers: Learning in a diffuse transnational network. *Cambridge Review of International Affairs, 19*(4), 555–569. http://dx.doi.org/10.1080/09557570601003205

Kaufman, H. (1975). The natural history of human organizations. *Administration & Society, 7*(2), 131–149. http://journals.sagepub.com/doi/pdf/10.1177/009539977500700201

Lai, C.-H. (2017). A study of emergent organizing and technological affordances after a natural disaster. *Online Information Review, 44*(4), 507–523. http://doi.org/10.1108/OIR-10-2015-0343

Lai, C.-H., Tao, C.-C., & Cheng, Y.-C. (2017). Modeling resource network relationships between response organizations and affected neighborhoods after a technological disaster. *VOLUNTAS, 28*(5), 2145–2175. https://doi.org/10.1007/s11266-017-9887-4

Lennon, J. T., & Jones, S. E. (2011). Microbial seed banks: The ecological and evolutionary implications of dormancy. *Nature Reviews Microbiology, 9*(2), 119–130. http://doi.org/10.1038/nrmicro2504

Leonardi, P. M. (2013). When does technology use enable network change in organizations? A comparative study of feature use and shared affordances. *MIS Quarterly, 37*(3), 749–775. http://doi.org/10.25300/MISQ/2013/37.3.04h

Levins, R. (1968). *Evolution in changing environments: some theoretical explorations* (No. 2). Priceton: Princeton University Press.

Mariotti, F., & Delbridge, R. (2012). Overcoming network overload and redundancy in interorganizational networks: The roles of potential and latent ties. *Organization Science, 23*(2), 511–528. http://dx.doi.org/10.1287/orsc.1100.0634

McPhee, R. D., & Corman, S. R. (1995). An activity-based theory of communication networks in organizations, applied to the case of a local church. *Communication Monographs, 62*(2), 132–151. http://dx.doi.org/10.1080/03637759509376353

Meier, P. (2014). Digital humanitarians: How big data is changing the face of humanitarian response. Boca Raton, FL: CRC Press.

Meyer, M. W. (1994). Turning evolution inside the organization. In J. A.C. Baum, & J. V. Singh (Eds.), *Evolutionary dynamics of organizations* (pp. 109–116). New York: Oxford University Press.

Mishal, S., & Rosenthal, M. (2005). Al Qaeda as a dune organization: Toward a typology of Islamic terrorist organizations. *Studies in Conflict & Terrorism, 28*(4), 275–293. http://dx.doi.org/10.1080/10576100590950165

Monge, P. R., Fulk, J., Kalman, M. E., Flanagin, A. J., Parnassa, C., & Rumsey, S. (1998). Production of collective action in alliance-based interorganizational communication and information systems. *Organization Science, 9*(3), 411–433. http://dx.doi.org/10.1287/orsc.9.3.411

Monge, P. R., Heiss, B. M., & Margolin, D. B. (2008). Communication network evolution in organizational communities. *Communication Theory, 18*(4), 449–477. http://doi.org/10.1111/j.1468-2885.2008.00330.x

Norris, F. H., Stevens, S. P., Pfefferbaum, B., Wyche, K. F., & Pfefferbaum, R. L. (2008). Community resilience as a metaphor, theory, set of capacities, and strategy for disaster readiness. *American Journal of Community Psychology, 41*(1), 127–150. http://dx.doi.org/10.1007/s10464-007-9156-6

Parker, M., Cheney, G., Fournier, V., & Land, C. (Eds.). (2014). *The Routledge companion to alternative organization.* New York: Routledge.

Pfefferbaum, R. L., Pfefferbaum, B., Van Horn, R. L., Klomp, R. W., Norris, F. H., & Reissman, D. B. (2013). The Communities Advancing Resilience Toolkit (CART): An intervention to build community resilience to disasters. *Journal of Public Health Management and Practice, 19*(3), 250–258. http://journals.lww.com/jphmp/Abstract/2013/05000/The_Communities_Advancing_Resilience_Toolkit.9.aspx

Pratt, M. G., & Foreman, P. O. (2000). Classifying managerial responses to multiple organizational identities. *Academy of Management Review, 25*(1), 18–42. http://dx.doi.org/10.2307/259261

Resnick, P. (2001). Beyond bowling together: Sociotechnical capital. *HCI in the New Millennium, 77*, 247–272

Rheingold, H. (2000). *The virtual community: Homesteading on the electronic frontier.* Cambridge, MA: MIT Press.

Saunders, S. L., & Kreps, G. A. (1987). The life history of the emergent organization in times of disaster. *The Journal of Applied Behavioral Science, 23*(4), 443–462. http://dx.doi.org/10.1177/002188638702300402

Scott, C. R. (2013). *Anonymous agencies, backstreet businesses, and covert collectives: Rethinking organizations in the 21st century.* Palo Alto, CA: Stanford University Press.

Sherrieb, K., Norris, F. H., & Galea, S. (2010). Measuring capacities for community resilience. *Social Indicators Research, 99*(2), 227–247. http://dx.doi.org/10.1007/s11205-010-9576-9

Stanley, S. M. (1981). *The new evolutionary timetable.* New York: Basic Books.

Starbird, K., & Palen, L. (2011). Voluntweeters: Self-organizing by digital volunteers in times of crisis. In *Proceedings of the 2011 Annual Conference on Human Factors in Computing Systems* (pp. 1071–1080). New York, NY: ACM.

Starkey, K., Barnatt, C. & Tempest, S. (2000). Beyond networks and hierarchies: Latent organizations in the UK television industry. *Organization Science, 11*(3), 299–305. http://dx.doi.org/10.1287/orsc.11.3.299.12500

Treem, J. W., & Leonardi, P. M. (2013). Social media use in organizations: Exploring the affordances of visibility, editability, persistence, and association. *Annals of the International Communication Association, 36*(1), 143–189. http://doi.org/10.1080/23808985.2013.11679130

van den Hooff, B. (2004). Electronic coordination and collective action: Use and effects of electronic calendaring and scheduling. *Information and Management, 42*(1), 103–114. https://doi.org/10.1016/j.im.2003.12.006

Walker, K. L., & Stohl, C. (2012). Communicating in a collaborating group: A longitudinal network analysis. *Communication Monographs, 79*(4), 448–474. http://doi.org/10.1080/03637751.2012.723810

Walther, J. B. (1996). Computer-mediated communication: Impersonal, interpersonal, and hyperpersonal interaction. *Communication Research, 23*(1), 3–43. http://doi.org/10.1177/009365096023001001

Weick, K. E. (1969). *The social psychology of organizing.* Menlo Park, CA: Addison-Wesley.

Weick, K. E. (1979). Cognitive processes in organizations. *Research in organizational behavior, 1*(1), 41–74.

Wenger, D. E., & James, T. F. (1994). The convergence of volunteers in a consensus crisis. In R. Dynes & K. Tierney (Eds.), *Disasters, collective behavior and social organization* (pp. 229–243). Newark: University of Delaware Press.

11

CONCLUSIONS AND FUTURE INTERDISCIPLINARY OPPORTUNITIES

Keri K. Stephens

The need to organize, handle, and cope with crises is not going away. The chapters in this book have provided a behind-the-scenes look at how practitioners, first responders, and citizens organize, use new media, and make sense of their experiences. Our goal is to provide an interdisciplinary look at crises with the hope of inspiring future research that capitalizes on synergies between academic fields. Historically, disaster research, in particular, has been an interdisciplinary field with roots in sociology (e.g., Kreps & Bosworth, 2008; Mileti & Sorensen, 1990), but other topics—for example, crisis communication—discussed here have been less interdisciplinary. With the ever-changing host of new media, an interdisciplinary approach that also considers the process of organizing is important. That is why I introduced the Model of Organizing in Times of Crisis (OTC, see Figure 11.1). This model reminds us that we can pursue research in the individual concept areas—i.e., new media, messages, public involvement, and crisis responders—but they are not independent of one another. Furthermore, the conceptual glue that binds these research areas is an understanding that organizing operates at and between all these spaces. This model provided a framework for this book, and as I survey the host of directions for future research suggested by the chapter authors, the model provides a guide.

While I have referred to the constellation of concepts addressed here as *times of crisis*, keep in mind that some concepts are more or less relevant during specific situations, such as disasters, emergencies, hazards, and crises. Here, I pull together these disparate contexts, and draw from emerging trends, to suggest

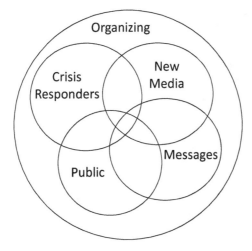

FIGURE 11.1   Model of Organizing in Times of Crisis (OTC)

where this research should go in the future. This research agenda is organized into three main categories:

1. Using interdisciplinary research to help practitioners and first responders.
2. Expanding beyond the new media prevalent today (e.g., Twitter), and the messages we send to better understand how organizing might happen in the future.
3. Trends in big data, artificial intelligence, and simulations that apply to organizing during times of crisis.

Interdisciplinary Research to Help Practitioners and First Responders

One contribution of our book is in its interdisciplinary approach; but pursuing this type of work is complex because the theories and concepts are spread among diverse journals, conference proceedings, and public science outlets. The brief list compiled here represents different publication outlets cited throughout this book and is evidence of the diverse knowledge bases that inform the chapters:

- *Communication Theory*
- *Communications of the ACM*
- *Handbook of Crisis Communication*
- *Handbook of Organizational Communication*
- *Handbook of Warnings*
- *Homeland Security and Emergency Management*

- *Information Systems for Crisis Response and Management. (ISCRAM)*
- *International Conference on Human Factors in Computing Systems (CHI)*
- *Journal of Contingencies and Crisis Management*
- *Journal of Loss and Trauma*
- *Journal of Management*
- *Journal of Safety Research*
- *Journal of Social and Personal Relationships*
- *Journal of Sociology*
- *Journal of Transportation Safety & Security*
- *Media Psychology*
- *Mobile Media & Communication*
- *Natural Hazards*
- *Planning Practice & Research*
- *Public Relations Review*

While these publication outlets span quite a range, the authors in this book did a nice job situating their own contributions and defining related terms that allowed them to narrow their approaches. Furthermore, there are important connections illustrated in this book that are useful in guiding future research.

Resilience Research and Crisis Responders

In the area of crises, much of the resilience research is focused on people and communities who experience a tragedy. Resilience research, as explained by Houston in Chapter 8, is inherently interdisciplinary since this work originated in psychology, has roots in medicine, and is now funded by the Department of Homeland Security (e.g., Norris, Stevens, Pfefferbaum, Wyche, & Pfefferbaum, 2008). Moving forward, a key opportunity is to better understand how to help crisis responders become more resilient. Other fields like healthcare are now addressing concerns like provider burnout (e.g., Panagioti et al., 2016), and they are finding that an emphasis on resilience research offers promise for supporting healthcare providers in times of crisis.

While emergency responders often receive some form of post-disaster counseling and/or group therapy, this is not necessarily the case for all crisis responders, especially volunteers (Laith, 2004). Considering that many crisis responders—volunteers and representative of official organizations—were in their positions before social media existed, this could be an added emotional burden. At the end of Chapter 1, Barrett and Posey discussed the lack of resources available for university crisis communicators to understand, cope with, and overcome burnout. There are many other communication job roles, like public information officers (PIOs) where handling and monitoring social media during times of crises has been a task added on to an existing job description (Hughes & Palen, 2012). While companies can train people in customer services and communication roles to handle new social media platforms and related tasks, this is not always the case with smaller and governmental organizations.

Volunteers who want to help during crises could also experience these challenges because they are not necessarily part of a system that provides them training and post-disaster counseling (Laith, 2004). Furthermore, when emergencies strike, local official organizations, like Health and Human Services, may not be able to help, and with the prominence of social media, citizens can provide one another with health and mental health information, whether it is correct or not (Stephens, Li, Robertson, Smith, & Murthy, 2018). Providing health and resilience resources to help these volunteers, as well as employees, cope with the emotion-intensive jobs they do is a natural next step for interdisciplinary research teams to pursue.

Apply Team Knowledge to Informal Crisis Teams

Jahn (Chapter 2) and Williams (Chapter 3) illustrate how teams coordinate, make sense of their experiences, and garner resources. It could be helpful to transfer some of those research findings to contexts of online and emergent groups like Lai (Chapter 10) and Hughes (Chapter 9) discussed. Hughes' notions of online convergence (Chapter 9) and Lai's elaboration of forms of dormant organizing (Chapter 10) provide solid ideas for how citizen teams function and disband in times of crisis. The challenges of ad hoc, emergent, and volunteer groups were identified over a decade ago (e.g., Kreps & Bosworth, 2008), and unfortunately, recent research finds these challenges have grown now that new media allows people myriad ways to communicate (Smith, Stephens, Robertson, Li, & Murthy, 2018; Whittaker, McLennan, & Handmer, 2015). While these groups often have good intentions and provide a helpful service, they also tend to suffer from unnecessary redundancies in their efforts, and confusion when communicating through a host of mobile apps and Web-based tools. By combining contemporary coordination research with the newer forms of online citizen groups, there could be fruitful ways to help informal groups be more effective.

Expanding beyond Media of Today and Better Understanding Message Interpretation

Several chapters have illustrated the wide range of messages that people can share and seek during crises and hazardous situations. Building off the work of Cacciatore, Kim, & Danzy (Chapter 5), it will be important to continue studying how misinformation originates, how it is sometimes deliberately shared through public social media (e.g., Starbird, Maddock, Orand, Achterman, & Mason, 2014), and how citizens and official organizations can discern truthful accounts. Perhaps borrowing from some of Ford's (Chapter 4) work on information seeking can help researchers develop a more complete understanding of the motivations behind how people seek crisis and emergency information. In fact, safety and emergency information seeking is a big issue for organizations, especially those tasked with operating with or around hazardous material. Ford (Chapter 4) sets a full research agenda as we expand our consideration of crises into the workplace context.

Incorporate Evacuation Knowledge with Maps and Organizing

One of the key values of having civil engineers contribute to this book is that social scientists can learn from the vast knowledge they have amassed on evacuations, transportation, and human needs like finding shelter. If we were to combine the knowledge of traffic and roadways shared by Rambha, Jafari, and Boyles (Chapter 7) with the research that Bean and Madden (Chapter 6) are doing concerning the inclusion of maps in brief communication messages, that could be a fruitful endeavor. Imagine being in an emergency and receiving a WEA that contains a geographically relevant evacuation map. Furthermore, that alert could contain a link to evacuation shelters and updates concerning which ones are closed and those that accept animals, important considerations when people evacuate their homes (White, Palen, & Anderson, 2014). That type of highly customized and immediately relevant information could be a persuasive way to provide enough helpful information to convince citizens to evacuate. This is just one example of the importance of these types of interdisciplinary collaborations, and there are many more worth exploring.

As informal groups continue organizing to help rescue and recovery efforts, understanding how they might use maps and evacuation routes could also be a helpful interdisciplinary exploration. In the years between 2015 and 2018, a number of mobile applications have been created with the goal of helping people visualize resources. For example, the Cajun Navy has many behind-the-scenes tools that they use to coordinate rescues (www. cajunnavyrelief.com/) and more publicly available maps from organizations like CrowdSource Rescue (https://crowdsourcerescue.com) provide citizens locations of emergency shelters, along with reports of people needing rescue. One nonprofit organization, also mentioned in Chapter 9, is Humanity Road (www.humanityroad.org/).They have been operating in the rescue and relief space since 2010 and they have experience using technology to connect people in a host of disasters worldwide. Some of these groups are also raising important issues concerning the privacy of people during a disaster, because having an address posted on a public map could leave disaster victims vulnerable as immediate emergencies subside. This is an area ripe for research.

Big Data, Artificial Intelligence, and Simulations

Organizations are filled with data, and there is already evidence that using the right analyses and visualizations can bring new meaning to this big data (Davenport & Harris, 2007), especially in a crisis and disaster context (e.g., O'Neal et al., 2018). Some of the tools being explored include artificial intelligence (AI), machine learning, and computer-based simulations. Computer algorithms are helping disaster relief organizations match online requests and recover lost property (Imran, Castillo, Lucas, Meier, & Vieweg, 2014).

One of the biggest challenges with these newer media platforms is finding what is needed (or the signal data) in a sea of noisy, unrelated data (Palen & Hughes, 2018; O'Neal et al., 2018; Qadir et al., 2016). Hughes (Chapter 9) shared several examples of nonprofit organizations, like Humanity Road, and machine-learning platforms, like Artificial Intelligence for Disaster Response (AIDR), that are actively engaged in using new media and data analyses techniques to improve disaster response and recovery (Imran, et al., 2014). One avenue ripe for further examination is harnessing big data to better prepare communities for disasters and emergencies that necessitate evacuations or shelter-in-place. We have decades of research that has identified who is likely to evacuate or follow officials' recommendations (see Rambha, Jafari, and Boyles' Chapter 7 for a comprehensive treatment of the past literature), now we need to use these findings to inform big data analyses.

Another research approach where past research can provide valuable insight is using computer simulations. Simulations—using computers to work through multiple outcome scenarios—are often used in fields like transportation and computer engineering because it is not feasible to test for different outcomes using other methods. This approach functions as a rapid experiment platform where the conditions (also called inputs, or variables that can affect change in a system) are rapidly changed to generate outcome data under a variety of inputs. Some agent-based modeling simulation work has been used to predict outcomes in times of crisis (e.g., Stephens, Jafari, Boyles, Ford, & Zhu, 2015). Stephens and colleagues were interested in knowing what would happen if ICTs were used to notify people they needed to evacuate. They had a complete set of transportation data, and by varying communication variables, they found that there is a rush to the road when ICTs are used to reach people quickly. The evacuation bottleneck is no longer stuck at the point where officials try to get evacuation messages to people, but instead, with ample communication channels available, the road system cannot handle that many vehicles, so the freeways become parking lots (Stephens et al., 2015).

There is limited published research using agent-based models in the social sciences, and that might be an ideal place to engage interdisciplinary research to solve larger disaster and emergency issues. The chapters in this book provide many ideas for how we can use past research to predict how people might behave. Those predictions are then used as inputs to generate a series of scenarios that can reveal the range of human behaviors. Notice that I have carefully stated the limits of simulations and agent-based modeling. The predicted outcomes are only as good as the decisions made when establishing input parameters. Furthermore, using this approach, actual human behavior is not measured or observed, therefore, it is helpful to consider simulations as producing a range of predicted outcomes.

Conclusion

Considering the rapid changes in new media, messages, crisis responders, and the public's involvement that have occurred in the past decade, it is important for scholars to pursue these research agendas. Some of the ideas shared are ready to be implemented and others might need to be addressed in small steps. For example, even if we have the capability of using AI, machine learning, and simulations to help plan and respond in times of crisis, there is no guarantee that first responders will be comfortable placing their lives in the hands of predictive analytics. As we continue studying new media in times of crises, we cannot forget that the decisions we make as scholars and practitioners have a profound impact on human lives and experiences.

References

Davenport, T. H., & Harris, J. G. (2007). *Competing on analytics: The new science of winning*. Boston, MA: Harvard Business School Press.

Hughes, A. L., & Palen, L. (2012). The evolving role of the public information officer: An examination of social media in emergency management. *Journal of Homeland Security and Emergency Management, 9*, Article 22. https://dx.doi.org/10.1515/1547-7355.1976

Imran, M., Castillo, C., Lucas, J., Meier, P., & Vieweg, S. (2014). AIDR: Artificial intelligence for disaster response. In *Proceedings of the Companion Publication of the 23rd International Conference on World Wide Web Companion* (pp. 159–162). Republic and Canton of Geneva, Switzerland: International World Wide Web Conferences Steering Committee. https://dx.doi.org/10.1145/2567948.2577034

Kreps, G. A., & Bosworth, S. L. (2008). Organizational adaptations to disaster. In H. Rodriguez, E. L. Quarantelli, & R. R. Dynes (Eds.), *Handbook of disaster research* (pp. 297–315). New York, NY: Springer-Verlag.

Laith, S. (2004). Averting a disaster within a disaster: The management of spontaneous volunteers following the 11 September 2001 attacks on the World Trade Center in New York. *Volunteer Action, 6*(2), 11–29.

Mileti, D. S., & Sorensen, J. H. (1990). *Communication of emergency public warnings: A social science perspective and state-of-the-art assessment*. Oakridge, TN: Oak Ridge National Laboratory.

Norris, F. H., Stevens, S. P., Pfefferbaum, B., Wyche, K. F., & Pfefferbaum, R. L. (2008). Community resilience as a metaphor, theory, set of capacities, and strategy for disaster readiness. *American Journal of Community Psychology, 41*, 127–150.

O'Neal, A., Rodgers, B., Segler, J., Murthy, D., Lakuduva, N., Johnson, M., & Stephens, K. K. (2018). Training an emergency-response image classifier on signal data. *Proceedings of the IEEE 17th International Conference on Machine Learning and Applications ICMLA*, Orlando, Florida.

Palen, L., & Hughes, A. L. (2018). Social media in disaster communication. In H. Rodriguez, W. Donner, & J. E. Trainor (Eds.), *Handbook of disaster research* (pp. 497–518). Cham, Switzerland: Springer. https://dx.doi.org/10.1007/978-3-319-63254-4_24

Panagioti, M., Panagopoulou, E., Bower, P., Lewith, G., Kontopantelis, E., Chew-Graham, C., Dawson, S., van Marwijk, H., Geraghty, K., & Esmail, A. (2016). Controlled interventions to reduce burnout in physicians: A systematic review and meta-analysis. *JAMA Internal Medicine*, e1–e11. https://doi.org./10.1001/jamainternmed.2016.7674

Qadir, J., Ali, A., Rasool, U. R., Zwitter, A., Sathiaseelan, A., & Crowcroft, J. (2016). Crisis analytics: Big data-driven crisis response. *Journal of International Humanitarian Action, 1*(2), 1–24. https://doi.org/10.1186/s41044-016-0002-4

Smith, W. R., Stephens, K. K., Robertson, B. W., Li. J., & Murthy, D. (2018). Social media in citizen-led disaster response: Rescuer roles, coordination challenges, and untapped potential. In K. Boersma & B. Tomaszewski (Eds.), *Proceedings of the 15th International ISCRAM Conference*. Rochester, NY. Retrieved from: http://idl.iscram.org/files/williamrsmith/2018/1586_WilliamR.Smith_etal2018.pdf

Starbird, K., Maddock, J., Orand, M., Achterman, P., & Mason, R. M. (2014). Rumors, false flags, and digital vigilantes: Misinformation on Twitter after the 2013 Boston Marathon bombing. *iConference*. https://doi.org/10.9776/14308

Stephens, K. K., Jafari, E., Boyles, S., Ford, J. L., Zhu, Y. (2015). Increasing evacuation communication through ICTs: An agent-based model demonstrating evacuation practices and the resulting traffic congestion in the rush to the road. *Journal of Homeland Security and Emergency Management, 12*, 497–528. https://doi.org/10.1515/jhsem-2014-0075

Stephens, K. K., Li, J., Robertson, B. W., Smith, W. R., & Murthy, D. (2018). Citizens communicating health information: Urging others in their community to seek help during a flood. In K. Boersma & B. Tomaszewski (Eds.), *Proceedings of the 15th International ISCRAM Conference*. Rochester, NY. Retrieved from: http://idl.iscram.org/files/kerikstephens/2018/1609_KeriK.Stephens_etal2018.pdf

White, J. I., Palen, L., & Anderson, K. M. (2014). *Digital mobilization in disaster response: The work & self-organization of on-line pet advocates in response to Hurricane Sandy*. Proceedings of CSCW, Baltimore, MD. https://doi.org/10.1145/25316022531633

Whittaker, J., McLennan, B., & Handmer, J. (2015). A review of informal volunteerism in emergencies and disasters: Definition, opportunities and challenges. *International Journal of Disaster Risk Reduction, 13*, 358–368. https://doi.org/10.1016/j.ijdrr.2015.07.010

ABOUT THE EDITOR AND AUTHORS

Keri K. Stephens (Ph.D., The University of Texas at Austin), is an Associate Professor in the Moody College of Communication, a faculty affiliate with the Center for Health Communication, and a faculty fellow with the Center for Health and Social Policy at the LBJ School of Public Affairs. Her research and teaching interests bring an organizational, organizing, and technology perspective to understanding complex contexts like crisis, emergency, disaster, and healthcare. She has over 60 peer-reviewed publications, and her most recent book is *Negotiating Control: Organizations and Mobile Communication* (Oxford University Press).

Ashley K. Barrett (Ph.D., The University of Texas at Austin, 2018) is an Assistant Professor in Health and Organizational Communication at Baylor University. Her research explores how organizations use new technology to communicate with their employees and other stakeholders during times of high uncertainty and high risk. Her work has been published in *Management Communication Quarterly*, *Health Communication*, *Human Communication Research*, and *Information People & Technology*.

Hamilton Bean, Ph.D., MBA, APR, is Associate Professor of Communication at the University of Colorado Denver, where he conducts research at the intersection of communication, organization, and security. Since 2005, he has been affiliated with the National Consortium for the Study of Terrorism and Responses to Terrorism (START), a U.S. Department of Homeland Security Center of Excellence.

Stephen D. Boyles (Ph.D., The University of Texas at Austin) is an Associate Professor in the Department of Civil, Architectural and Environmental Engineering at The University of Texas at Austin.

Michael A. Cacciatore (Ph.D., University of Wisconsin-Madison) is an Assistant Professor in the Department of Advertising and Public Relations in Grady College at the University of Georgia. His research focuses on media coverage of and opinion formation for science and risk topics.

Dasia Danzy is an undergraduate student majoring in Public Relations in Grady College at the University of Georgia. Danzy is interested in strategic sports communication.

Jessica L. Ford (Ph.D., The University of Texas at Austin) is an Assistant Professor of Health Communication at Baylor University. Her research examines organizational disruptions that affect the health and safety of organizational members. Dr. Ford's research often investigates the implications of technology use during emergency events, such as evacuations, school shootings, and workplace injuries.

J. Brian Houston, Ph.D., is Chair and Associate Professor in the Department of Communication and Director of the Disaster and Community Crisis Center (dcc.missouri.edu) at the University of Missouri. Houston's research focuses on communication at all phases of disasters and on the mental health effects and political consequences of community crises.

Amanda Lee Hughes (Ph.D., University of Colorado Boulder) is an Assistant Professor of Information Technology and Cybersecurity at Brigham Young University. Her research interests span human–computer interaction, computer-supported cooperative work, social computing, software engineering, and disaster studies. Her current work investigates the use of social media during crises and mass emergencies, with particular attention to how they affect emergency response organizations.

Ehsan Jafari (Ph.D.) is currently a Senior System Engineer at Optym, and contributed to this chapter while a doctoral student at The University of Texas at Austin.

Jody L. S. Jahn (Ph.D., University of California, Santa Barbara) is an Assistant Professor at University of Colorado at Boulder. Her research uses mixed methods to examine how members of hazardous organizations communicate to negotiate action, and how members interface with organizational safety policies and documents. Recent research examines wildland firefighting workgroups. Her research appears in *Management Communication Quarterly, Journal of Applied Communication Research, Communication Monographs*, and other journals.

Sungsu Kim is a Ph.D. candidate in Grady College at the University of Georgia. His research focuses on strategic crisis communications.

Chih-Hui Lai (Ph.D., Rutgers University, 2012) is an Associate Professor in the Department of Communication and Technology at National Chiao Tung University, Taiwan. Her research investigates how individuals, groups, and organizations use multiple media modalities to communicate and coordinate in the face of disasters, and how relationships evolve or emerge through these processes. Her works have been published in top-tier communication journals, including *Journal of Communication, Communication Research, Journal of Computer-Mediated Communication, Human Communication Research*, and many others.

Stephanie Madden, Ph.D., is Assistant Professor in the Department of Journalism and Strategic Media at the University of Memphis. Her research explores the intersections among public relations, activism, risk/crisis communication, and social media. Previously, Dr. Madden was a full-time communication researcher at the National Consortium for the Study of Terrorism and Responses to Terrorism (START), a U.S. Department of Homeland Security Center of Excellence.

Cindy Posey, M.A., is Director of Internal Communications, and the former Director of Internal and Campus Safety Communications and Public Information Officer, at The University of Texas at Austin. She has more than 30 years of experience in organizational and crisis communication, has taught as an adjunct instructor at the University of Central Florida and Southwestern University, and is a certified executive coach.

Tarun Rambha (Ph.D.) is an Assistant Professor in the Department of Civil Engineering at the Indian Institute of Science Bengaluru. He conducts research in transportation network modeling and optimization.

Elizabeth A. Williams (Ph.D., Purdue University) is an Associate Professor in the Department of Communication Studies at Colorado State University. Her research and teaching is at the intersection of organizational and health communication research. She is interested in how multiteam systems and high reliability organizations learn and recover from failures and how those processes influence the health and safety of organizational members.

INDEX

For Product Safety Concerns and Information please contact our EU
representative GPSR@taylorandfrancis.com
Taylor & Francis Verlag GmbH, Kaufingerstraße 24, 80331 München, Germany